List of Tables

Problem Solver for

FINITE MATHEMATICS *and* CALCULUS

The Prindle, Weber & Schmidt Series in Mathematics

Althoen and Bumcrot, *Introduction to Discrete Mathematics*
Brown and Sherbert, *Introductory Linear Algebra with Applications*
Boye, Kavanaugh, and Williams, *Elementary Algebra*
Boye, Kavanaugh, and Williams, *Intermediate Algebra*
Buchthal and Cameron, *Modern Abstract Algebra*
Burden and Faires, *Numerical Analysis, Fourth Edition*
Cass and O'Connor, *Fundamentals with Elements of Algebra*
Cullen, *Linear Algebra and Differential Equations, Second Edition*
Cullen, *Mathematics for the Biosciences*
Dick and Patton, *The Oregon State University Calculus Curriculum Project*
Eves, *In Mathematical Circles*
Eves, *Mathematical Circles Adieu*
Eves, *Mathematical Circles Revisited*
Eves, *Mathematical Circles Squared*
Eves, *Return to Mathematical Circles*
Fletcher and Patty, *Foundations of Higher Mathematics*
Gantner and Gantner, *Trigonometry*
Geltner and Peterson, *Geometry for College Students, Second Edition*
Gilbert and Gilbert, *Elements of Modern Algebra, Second Edition*
Gobran, *Beginning Algebra, Fifth Edition*
Gobran, *Intermediate Algebra, Fourth Edition*
Gordon, *Calculus and the Computer*
Hall, *Algebra for College Students*
Hall and Bennett, *College Algebra with Applications, Second Edition*
Hartfiel and Hobbs, *Elementary Linear Algebra*
Kaufmann, *Algebra for College Students, Third Edition*
Kaufmann, *Algebra with Trigonometry for College Students, Second Edition*
Kaufmann, *College Algebra, Second Edition*
Kaufmann, *College Algebra and Trigonometry, Second Edition*
Kaufmann, *Elementary Algebra for College Students, Third Edition*
Kaufmann, *Intermediate Algebra for College Students, Third Edition*
Kaufmann, *Precalculus, Second Edition*
Kaufmann, *Trigonometry*
Laufer, *Discrete Mathematics and Applied Modern Algebra*
Nicholson, *Elementary Linear Algebra with Applications, Second Edition*
Patty, Fletcher, and Hoyle, *Foundations of Discrete Mathematics*
Pence, *Calculus Activities for Graphic Calculators*
Powers, *Elementary Differential Equations*
Powers, *Elementary Differential Equations with Boundary-Value Problems*
Powers, *Elementary Differential Equations with Linear Algebra*
Proga, *Arithmetic and Algebra, Second Edition*
Proga, *Basic Mathematics, Third Edition*
Rice and Strange, *Plane Trigonometry, Fifth Edition*
Schelin and Bange, *Mathematical Analysis for Business and Economics, Second Edition*
Strnad, *Introductory Algebra*

Swokowski, *Algebra and Trigonometry with Analytic Geometry, Seventh Edition*
Swokowski, *Calculus with Analytic Geometry, Fifth Edition*
Swokowski, *Calculus with Analytic Geometry, Fourth Edition (Late Trigonometry)*
Swokowski, *Fundamentals of Algebra and Trigonometry, Seventh Edition*
Swokowski, *Fundamentals of College Algebra, Seventh Edition*
Swokowski, *Fundamentals of Trigonometry, Seventh Edition*
Swokowski, *Precalculus: Functions and Graphs, Sixth Edition*
Tan, *Applied Calculus, Second Edition*
Tan, *Applied Finite Mathematics, Third Edition*
Tan, *Calculus for the Managerial, Life, and Social Sciences, Second Edition*
Tan, *College Mathematics, Second Edition*
Trim, *Applied Partial Differential Equations*
Venit and Bishop, *Elementary Linear Algebra, Third Edition*
Venit and Bishop, *Elementary Linear Algebra, Alternate Second Edition*
Wiggins, *Problem Solver for Finite Mathematics and Calculus*
Willard, *Calculus and Its Applications, Second Edition*
Wood and Capell, *Arithmetic*
Wood, Capell, and Hall, *Developmental Mathematics, Fourth Edition*
Wood and Capell, *Intermediate Algebra*
Zill, *A First Course in Differential Equations with Applications, Fourth Edition*
Zill, *Calculus with Analytic Geometry, Second Edition*
Zill, *Differential Equations with Boundary-Value Problems, Second Edition*

The Prindle, Weber & Schmidt Series in Advanced Mathematics

Brabenec, *Introduction to Real Analysis*
Ehrlich, *Fundamental Concepts of Abstract Algebra*
Eves, *Foundations and Fundamental Concepts of Mathematics, Third Edition*
Keisler, *Elementary Calculus: An Infinitesimal Approach, Second Edition*
Kirkwood, *An Introduction to Real Analysis*
Ruckle, *Modern Analysis: Measure Theory and Functional Analysis with Applications*

Problem Solver for

FINITE MATHEMATICS *and* CALCULUS

KENNETH L. WIGGINS

Walla Walla College

PWS-KENT PUBLISHING COMPANY

Boston

PWS-KENT
Publishing Company

20 Park Plaza
Boston, Massachusetts 02116

PWS-KENT Publishing Company is a division of Wadsworth, Inc.

Library of Congress Cataloging-in-Publication Data

Problem solver for finite mathematics and calculus/
Wiggins, Kenneth L.

p. cm.
Includes index.
ISBN 0-534-92439-5
1. Mathematics. 2. Calculus. 3. Mathematics- Data processing
4. Calculus- Data Processing. 5. Problem solver (Computer program)
I. Title
QA39.2.W54 1991
510--dc20 90-40928
CIP

Printed in the United States of America

90 91 92 93 94 - 10 9 8 7 6 5 4 3 2 1

Sponsoring Editor Steve Quigley
Production Editors Helen Walden/Cathy Lozyniak
Text and Art Preparation Kenneth L. Wiggins
Cover Design Cathy Lozyniak
Manufacturing Coordinator Lisa Flanagan
Printing and Binding Malloy Lithographing, Inc.
Cover Printer Henry Sawyer Company

Contents

Preface

Computers are wonderful tools for mathematics students. This book and *Problem Solver*, the accompanying computer program, work together to bring the power of the computer to students. This material is primarily intended for students enrolled in finite mathematics and applied calculus courses. In addition, it can be used in precalculus courses or in the first two terms of a standard calculus sequence. It can serve most effectively as a supplement to a standard textbook by providing students with a wide variety of computer activities. No experience in programming or using computers is required. The examples and applications included in this book range from some that are very simple to others that are quite sophisticated. They fall into two groups: those that emphasize the practical side of mathematics and those that allow the student to use the computer as a laboratory in which to explore theoretical concepts.

To the Student

The software packaged with this book fills an important need in the problem solving process by removing the burden of long and tedious calculations. Ease-of-use was the preeminent concern during the development of this program. To help you learn to use the software, each section of this book includes a special example titled *Problem Solver Example*. These examples are tutorials; they give step-by-step instructions for using the program to help you solve problems. An extensive on-line help system is also available while you are running Problem Solver. If you need more detailed help, see Appendix C. The complete reference guide for Problem Solver contained in this appendix even includes a collection of questions most commonly asked by students along with answers to these questions.

The exercises are designed to help you understand mathematical concepts more clearly through numerical experimentation and to provide practice in setting up and solving interesting and sometimes challenging applied mathematical problems. When you approach applied problems, consider the following points:

1. If possible, draw a picture illustrating the important points in the problem.

2. Carefully define the variables that you intend to use.

3. Write down any relationships that exist among the variables that you have defined.

4. Check to see exactly what information is being requested. Can this be expressed in terms of your variables?

5. Have you previously worked a similar problem?

6. It may be helpful to convert the problem to an easier problem. Solving the easier problem may provide insight into the more difficult problem.

7. Solve the problem. Is your answer reasonable?

To the Instructor

By furnishing both the software and suggested computer activities, this material provides a practical means of introducing computer usage into a variety of mathematics courses. The software is so easy to learn and use that very little class time need be devoted to the details of using it. The purpose of this package is fourfold:

1. The program can be used as a laboratory where students can explore mathematical concepts experimentally. For example, limits and derivatives can be explored through tables of functional values and graphs. Infinite series can be studied by constructing sequences of partial sums. Statistical ideas can be studied through simulation. Random data sets from various distributions can be generated and studied.

2. The examples and exercises in standard textbooks must be limited to problems that can reasonably be worked by hand. The Problem Solver software brings much larger and more realistic problems within the grasp of students.

3. For many algorithms, intermediate calculations can be displayed to promote a better understanding of the methods used.

4. A projection unit for an overhead display or a large display screen and this program make an excellent tool for in-class demonstrations. These demonstrations can be prepared in advance, saved in disk files, and then loaded and run in class. This makes it easy for instructors to answer "what if" questions posed by students.

How to Use This Book

The following core material includes information on running the software, solving equations, and graphing functions. These sections should be studied by all students:

- Chapter 1
- Chapter 2, sections 2.1 and 2.2
- Chapter 3, section 3.1

Additional material from chapters 2, 3, and 4 can be selected as needed for courses in finite mathematics. For precalculus students, additional material from chapters 2, 3, and 5 can be used. Calculus students should study the following sections in sequence:

- Chapter 5, sections 5.1, 5.2, and 5.3
- Chapter 6, sections 6.1 and 6.2
- Chapter 7, sections 7.1 and 7.2
- Chapter 8, section 8.1 and 8.2

Sections 5.4 and 7.3 are optional.

Each section contains exercises of varying levels of difficulty. Most are very straightforward, but a few are more challenging. These are annotated with $\star$. A few very challenging exercises are marked $\star\star$. These could be used as long-term or group projects. Exercises requiring trigonometry are marked $\triangle$.

Preparing Written Reports

The results of calculations done by Problem Solver are always displayed on the video monitor. In addition, this output can be printed immediately or sent to a disk file for later printing. Both the format and the completeness of these reports can be controlled. For those courses emphasizing writing skills in mathematics, instructors may choose to ask their students to direct the program output to a disk file. This output file can then be loaded into a word processor and comments can be added as needed. The results can then be printed from within the word processor.

Acknowledgements

I would like to thank those who reviewed this work and provided numerous helpful suggestions.

Dean R. Brown, Youngstown State University
Dennis C. Ebersole, Northampton Community College
B. A. Fusaro, Salisbury State College
Warren B. Gordon, Baruch College
James W. Hall, Parkland College
Kenneth E. King, Indiana Vocational Technical College
L. Carl Leinbach, Gettysburg College

Thanks are also due to

Ed Parker for encouraging me to begin this project.
My students who willingly used the program during its development.

Thomas Thompson and Ward Soper for using this program in their courses during its development and for providing valuable suggestions.

Thomas Thompson for his help with LaTeX, the typesetting program used to prepare this manuscript.

Gladys Werth for carefully proofreading the manuscript.

Bob Clark, Terry Fritz, and Graham Wheeler for working the odd-numbered problems.

Steve Quigley and the editorial staff at PWS-KENT for their expertise and encouragement.

I owe much to my wife JoAnn for her help with the graphics in this book, for proofreading, and for her enthusiastic support and encouragement throughout this project.

klw

Chapter 1

Using Problem Solver

1.1 Introduction

The program provided with this book is called *Problem Solver for Finite Mathematics and Calculus* or just *Problem Solver*. It acts as a super calculator to remove the tedium from many of the problems that you will be solving. This means that you will be able to focus your attention on setting up and analyzing mathematical problems rather than on carrying out extended calculations. It also means that much larger and more interesting problems will be within your grasp. Problem Solver is an easy-to-use program with the following features:

- Functions can be graphed easily and quickly. You can even graph up to 10 functions on the same coordinate system. To make your graphs more attractive, you can add headings and other annotation.
- Single nonlinear equations can be solved using a variety of methods.
- Calculus students can study limits, derivatives, and integrals using both tables and graphs.
- The available matrix operations include solving systems of linear equations, finding inverses of matrices, transforming matrices to row-reduced form, and finding sums and products of matrices.
- Permutations and combinations can be calculated.
- You can find the line of best fit to a set of data points.
- You can compute probabilities using the binomial or normal distributions. You can even generate random data sets from several distributions.

- Linear programming problems, matrices, and data for statistical analysis can be entered using a spreadsheet-style entry screen. This allows you to correct errors by simply moving the cursor to the incorrect entry and typing the correct entry. You can even enter expressions when numbers are requested. If you want to enter 7 divided by 8, there is no need to convert 7/8 to a decimal. Just enter 7/8, and the program will interpret your entry correctly. All of the standard scientific functions are available.

- Problems can be saved on your disk for later use.

- For many of the algorithms used, you can view the intermediate calculations.

- Program output is always displayed on the screen. In addition, it can be sent to a printer or even stored in a disk file. Output stored in a disk file can be edited using a word processing program.

- You can enter functions either as formulas or tables of values.

In addition to the short introduction to Problem Solver given in this chapter, you will find examples entitled "Problem Solver Example" in every section of this book. These examples are tutorial in nature. They include all of the details needed for getting Problem Solver to help you solve problems. For more in-depth information on the program, see the reference manual in Appendix C.

1.2 Getting Started

1.2.1 What You Need

The Problem Solver can be run on an IBM-compatible computer with the following equipment:

- At least 512KB of random access memory (RAM)

- A graphics monitor

- DOS version 3.0 or newer

For paper copies of your results, you will also need a printer. Any printer will work for text output, but for graphs to be printed, your printer must be compatible with Epson graphics printers.

1.2.2 Installing Problem Solver on Floppy Disks

This section involves the use of DOS commands for formatting a floppy disk, copying files, and setting up subdirectories. If you need more detailed help than this section provides, see Appendix B for a brief introduction to both personal computers and DOS.

Disks with a capacity of only 360 KB do not have room for both Problem Solver and the DOS system files. This means that you can make a boot disk for Problem Solver only if you are using higher capacity disks. For 360 KB disks you will need to have a separate boot disk. You can use your DOS disk for this purpose.

Format a floppy disk using the following steps:

1. Place your DOS disk in drive A (the left or top drive if you have two drives). When you see the `A>` prompt on the screen, type one of the following commands: The first command is for 360 KB disks, and the second is for higher capacity disks. Type

   ```
   FORMAT A:
   ```

   or

   ```
   FORMAT A:/S
   ```

   and then press the Enter key.

2. Remove the DOS disk from drive A and replace it with a blank disk.

3. Follow the instructions on the screen.

4. When the formatting is complete, remove the disk and label it *Problem Solver Program Work Disk.*

5. If you are using 360KB floppy disks, format a second floppy disk with the command

   ```
   FORMAT A:
   ```

6. Label your second disk *Problem Solver Help Work Disk.*

After the formatting is complete, you can copy the Problem Solver program files onto the disk that you labeled *Problem Solver Program Work Disk* using the following steps:

1. Place the Problem Solver Program Disk in drive A and type the command

   ```
   COPY A:*.* B
   ```

   Wait a moment before pressing the Enter key.

2. If you have two disk drives, place the disk that you labeled *Problem Solver Program Work Disk* in drive B and press the Enter key. If you have only one drive, press the Enter key. In this case, DOS will refer to the Problem Solver Program Work Disk as the *diskette for drive B* and to the Problem Solver Program Disk as the *diskette for drive A*. You will be instructed when to change disks.

3. After the copying is complete, repeat the above instructions to copy the files on the Problem Solver Help Disk. If you are using 360KB floppy disks, copy these files onto your Problem Solver Help Work Disk. If you are using higher capacity disks, you can copy these files onto your Problem Solver Program Work Disk.

To finish the installation, you need to place one or two additional files on the Problem Solver Program Work Disk. One of these files is called CONFIG.SYS, and it need contain only the single line

```
FILES = 20
```

You can easily create this file using the following steps:

1. If you are using 360 KB disks, put your boot disk or DOS disk in drive A. If you are using higher capacity disks, place the Problem Solver Program Work Disk in drive A.

2. Type the command

   ```
   COPY CON CONFIG.SYS
   ```

   and press the Enter key.

3. Type

   ```
   FILES = 20
   ```

   and press the F6 key or Ctrl-Z.

4. Press the Enter key.

Now remove your disk from drive A. If you have two disk drives, place it in drive B.

The other required file is called GRAPHICS.COM, and it is contained on one of your DOS disks. Find the disk with this file and place it in drive A. Type the command

```
COPY A:GRAPHICS.COM B:
```

and press the Enter key. This file should be placed on the Problem Solver Program Work Disk regardless of capacity of the disks.

1.2.3 Installing Problem Solver on a Hard Disk

Use the following steps to install Problem Solver on your hard disk:

1. Switch to drive C by typing

   ```
   C:
   ```

   and pressing the Enter key.

2. Switch to the root directory on drive C by typing

   ```
   CD \
   ```

   and pressing the Enter key.

3. Make a subdirectory called TPS by typing

   ```
   MDIR TPS
   ```

   and pressing the Enter key.

4. Change to your new subdirectory using the command

   ```
   CDIR \TPS
   ```

   Press the Enter key.

5. Place the Problem Solver Program disk in drive A and type the command

   ```
   COPY A:*.*
   ```

   Press the Enter key.

6. Place the Problem Solver Help disk in drive A and type the command

   ```
   COPY A:*.*
   ```

   Press the Enter key.

Now change to the root directory by typing

```
CDIR \
```

and pressing the Enter key. If you have a file called CONFIG.SYS in the root directory, it should contain the line

```
FILES = 20
```

If it has this line with a number not less than 20, that is fine. Otherwise this file must be changed. Use a word processor to make this modification. If no such file exists, you can create one using the commands given above for floppy disks.

Your AUTOEXEC.BAT file must contain a PATH statement, which includes the path to the subdirectory containing the DOS programs. You should also include a path to the Problem Solver directory. For example, if your DOS files are contained in the directory `C:\DOS`, you need the line

```
PATH C:\TPS;C:\DOS
```

See Appendix B or your DOS manual for more information on the PATH command and AUTOEXEC.BAT files.

1.3 Starting Problem Solver

Use the following steps to start Problem Solver on a floppy disk:

1. If your computer is not turned on, place your boot disk or DOS disk in drive A if you are using 360 KB disks. For higher capacity disks, place your program work disk in drive A. Then turn your computer on. If your computer is already running place your Problem Solver Program Work Disk in drive A.

2. Type

   ```
   PS
   ```

   and press the Enter key.

3. You should now see the opening screen for Problem Solver. Press the space bar to advance to the main menu.

4. If you are using 360KB disks, remove the Problem Solver Program Work Disk from the drive and replace it with the Problem Solver Help Work Disk.

To run Problem Solver on a system equipped with a hard disk, type the command

```
PS
```

As with all DOS commands, press the Enter key after typing the command.

Functions	Matrices	Statistics	Options

Figure 1.1: Main Menu

1.4 Menus, Help, and Editing

After loading Problem Solver, you should see the main menu — the horizontal list of choices across the top of the screen shown in Figure 1.1. Your choice is indicated by the placement of the highlighted bar. You can change your selection in two ways:

- Use the arrow keys on the right side of the keyboard to move the highlighted bar to the selection that you want. When the bar is in the correct position, press the Enter key.

- Press the key corresponding to the highlighted letter in the option that you wish to select. For example, to select Statistics, you could press the *s* key.

Practice using the menus now. Move the cursor to "Options," and press the Enter key. From now on we shall refer to this action as "Selecting Options." Many of the Problem Solver menu options contain data that you can change. To see how this works, select Format for text output, and then select Decimal places in matrices. You will be given the opportunity to change this value. Type 4, press the Enter key, and then press the Esc key to return to the Options menu.

Other menu items involve yes-or-no responses. The first item on the Options menu falls into this category. If you are studying the techniques used by Problem Solver and if you wish to see the details of the calculations, select the first item to change the `no` to `yes`. Leave the Options menu by pressing the Esc key. When the Options menu closes, the active menu will again be the main menu. In summary, the menus in Problem Solver are used for two purposes:

- They allow the user to control what Problem Solver will do next. For example, the user can control whether Problem Solver draws a graph or solves an equation.

- In many situations data is entered into the menus by the user. We saw this in the Options menu. The choices built into Problem Solver are called *default* choices. To see how this works, select Options again. Then select Intermediate calculations displayed. Problem Solver will guess what your response will be. If this default choice is correct, press the Enter key. Otherwise press either the *y* key or the *n* key.

1.4.1 Using the Esc Key

The Esc key is used in three ways:

- Pressing the Esc key will take you out of the current menu and return you to the previous menu.
- If the current menu is the main menu, pressing the Esc key will allow you to exit the program.
- At times you may accidentally find yourself in a screen requesting data. Just press the Esc key to return to the previous screen.

1.4.2 Online Help

Help is as close as your F1 key. If you are not viewing the main menu, press the Esc key until the main menu is all that is displayed on the screen. Press the F1 key, and read all of the help screens that are available from the main menu. Page through these screens using the PgUp and PgDn keys on the right side of your keyboard. When you have finished viewing the help for the opening menu, press the Esc key to leave the help system.

Each menu has its own help screens. Select [Options] again, and press F1 to see the help available for this menu. For help on a particular topic, move to the appropriate menu and press the F1 key.

1.4.3 The Line Editor

Problem Solver uses two editors. The line editor is used for entering numbers and text, and the matrix editor is used for entering tables of numbers. The line editor is used each time you are asked to enter numbers or text into Problem Solver. The following commands are available:

- The arrow keys move the cursor to the left or right.
- The Home and End keys move the cursor to the beginning and end of your line, respectively.
- The Del key deletes the character at the cursor position.
- The Backspace key deletes the character to the left of the cursor.
- The Esc key is used if you wish to abandon your entry and use the original default choice.
- The Ins key is used to switch back and forth between the insert mode and the typeover mode.

```
C:\PS\DATA\*.DAT
```

Figure 1.2: Default Directory Window

- Press the Enter key when you have finished typing your input. This enters your data into Problem Solver.

Should you forget any of the above items, press the F1 key for help when you are in the line editor. The help screens will list these commands for you.

1.4.4 The Matrix Editor

The matrix editor is used for entering data for systems of equations, sets of points to plot, linear programming problems, and statistical data. Whenever the data needed is tabular in nature, the matrix editor is used.

This editor resembles a spreadsheet program. The screen is divided into a rectangular array of cells. To enter data into a cell, first move the cursor to that cell using the arrow keys; the PgUp and PgDn keys; or the right and left arrow keys in combination with the Ctrl key. Then type your new entry. As you type, you will be using the line editor, and all of the commands listed above are at your disposal. After you have finished typing an entry, press the Enter key to record your input. A detailed list of the commands available in the matrix editor is available in Appendix C. You can also see a list of these commands by pressing the F1 key while you are viewing the matrix editor.

1.4.5 Mathematical Expressions

Mathematical expressions can be entered into Problem Solver whenever numerical data is requested. If you want to enter the square root of 5, enter it as `SQRT(5)`. The available scientific functions are listed in Appendix C and in the Problem Solver help screens.

When you enter expressions, use `*` to indicate multiplication, and use `^` for exponents. Also remember that the arguments of all functions must be enclosed in parentheses. If you wish to enter the expression $3^2 + 4sin\frac{\pi}{4}$, you should type `3^2 + 4*sin(pi/4)`. You may use either upper or lower case letters in your input.

1.5 Using Files

Problem Solver allows you to save problems in files for future use. Many of the menus contain a save command. To save your work, select this option. You will be asked to enter a file name for your problem.

BALL.DAT	TIME.DAT
BISECT.DAT	URENT.DAT
BRIDGE.DAT	ZIP.DAT
MATRIX.DAT	
SAMPLE.DAT	

Figure 1.3: File Screen

To retrieve a data file, select the load command from the menu related to your problem. The first window to appear contains the name of the default directory. See Figure 1.2. You can edit the directory name if you wish to change directories, or if you remember your file name, you can type the file name in this window. Most likely, you will want to press the Enter key in order to see a list of your files like the list shown in Figure 1.3. When the file list appears on the screen, use the arrow keys to move the cursor to the file that you wish to load. Then press the Enter key to make your selection. If you need help in this screen, press the F1 key.

1.6 Using a Printer

To print your output, select Options from the main menu, and then from the Options menu select Output device. From the next menu, select Printer to send your output directly to a printer. Another alternative is to select File. Then all of your text output will be sent to a disk file, which can be loaded into a word processor for final editing and printing. Graphs can not be sent to files.

To control the margins and other formatting details for your printed text output, select Format for output from the Options menu.

In order to print graphs, your printer must be compatible with Epson printers. In addition, graphs displayed on monochrome Hercules-type screens may not print properly.

If you have the proper hardware, the printed graphs should look like your screen displays. If you have any difficulty printing graphs, try starting Problem Solver with the command

```
PS CGAHI
```

For EGA or VGA monitors, this will reduce the resolution of the graphs in the screen displays, but it will make the printing feature more compatible with older versions of DOS. With 4.0 and later versions of DOS this should not be necessary.

Chapter 2

Functions, Models, and Graphs I

2.1 Polynomials and Their Graphs

Polynomial functions are defined by equations of the form

$$f(x) = a_n x^n + a_{n-1} x^{n-1} + \cdots + a_1 x + a_0$$

where n is a nonnegative integer. If a_n is not zero, we say that the degree of f is n. The functions

$$g(x) = 2x^2 - x - 2,$$

$$p(x) = \frac{1}{8}x^3 - \frac{1}{2}x,$$

$$f(x) = 3x + 6,$$

and

$$h(x) = 7$$

are examples of polynomial functions with degrees 2, 3, 1, and 0, respectively. The graphs of polynomials are smooth curves with no angular corners and no breaks. The graph of a second degree polynomial function such as g is called a parabola. A first degree polynomial function such as f is called a linear function, and its graph is a line.

2.1.1 Problem Solver Example

We shall graph the equation $y = x^2 - 4$ for $x \in [-6, 6]$ and $y \in [-5, 5]$ using the following steps:

1. Start Problem Solver and select Functions from the main menu.

2. Then select Edit/enter functions.

3. Select the first function.

4. You should now be in the function editor. Type the expression `x^2 - 4` and then press the Enter key.

5. Press the Esc key to return to the functions menu.

6. Select Graph functions and from the next menu select Function selection. You should see a list of functions including the one that you entered.

7. Select the function that you entered. You will be asked if you wish to graph the function. Type `Y`, or, since Y is offered as the default response, just press the Enter key. Then press the Esc key to leave this menu.

8. Select Draw graph. When you are finished viewing the graph, press the Esc key to return to the graph functions menu. Did you notice that the graph was drawn on the interval [-5,5]? This interval needs to be changed.

9. Select Change graph specifications and then from the list of functions, select the one that you entered.

10. Select Minimum x-value to graph and enter `-6`. Notice that the default value for the minimum x-coordinate on the axis was -5. Since you requested a graph for x as small as -6, Problem Solver automatically changed the minimum x-coordinate on the graph from -5 to -6.

11. Select Maximum x-value to graph and enter `6`. Since the default maximum x-value on the axis was 5, and since you requested a graph for x as large as 6, Problem Solver automatically changed the maximum x-coordinate on the axis to 6.

12. Select Minimum y-coordinate on axis and enter `-5`. Then select Maximum y-coordinate on axis and enter `5`.

13. You can enhance the appearance of your graph by adding labels for your axes and headings for your graph. Experiment with these options. You might enter `x` as the label for the x-axis, `y` as the label for the y-axis, and `y = x^2 - 4` as the main heading for the graph. After you are finished, press the Esc key.

    Important note: When graphing several functions on the same coordinate system, you should enter a title for each function. This will help you identify the functions in the graph.

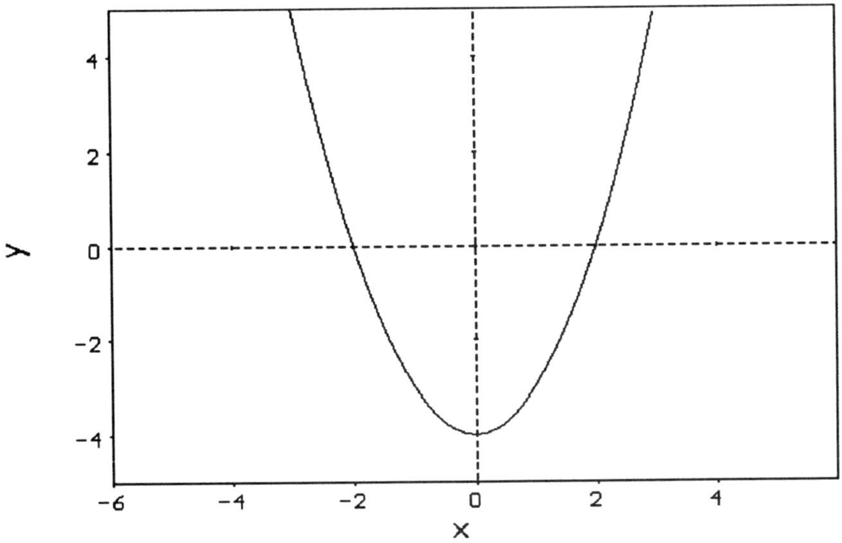

Figure 2.1: Graph of the $y = x^2 - 4$

14. Look at your graph again by selecting [Draw Graph]. Your graph should look like the graph shown in Figure 2.1.

2.1.2 Problem Solver Example

The line with equation $y = 2x - 1$ passes through the points (1,1) and (3,5). We shall graph the line and the two points on the same coordinate system for $x \in [0, 4]$, and $y \in [-1, 9]$.

1. Start Problem Solver and select [Functions] from the main menu.

2. Then select [Edit/enter functions].

3. Select the first function.

4. You should now be in the function editor. Type the expression `2*x - 1` and then press the Enter key.

5. Move the cursor down to the first undefined function (probably the second function) and press the Enter key.

6. In order to enter the two points, select [Table of values] from the next menu. When asked the number of points, type 2 and press the Enter key.

7. When asked for a title for your table, enter `Table 1` or any other title that you like. This title will appear in the list of functions so that you can select it for graphing or editing.

8. The next screen is the matrix editor. You will enter the ordered pairs using this screen. The first ordered pair goes in the first line, and the second ordered pair goes in the second line. Type `1` and press the Enter key. This should place `1` in the upper left cell.

9. When you pressed the Enter key, the cursor moved one position to the right for your next entry. Type `1` and press the Enter key. The cursor should move to the left side of the next line. Enter 3 into this cell and 5 into the lower right cell.

10. Press the Esc key to leave the matrix editor.

11. Press the Esc key again to return to the functions menu.

12. Select [Graph functions] and from the next menu select [Function selection]. You should see a list of functions including the two that you entered.

13. Select the functions that you entered. You will be asked if you wish to graph these functions. Type `Y`, or, since `Y` is offered as the default response, just press the Enter key. Do this for each of your two functions. Then press the Esc key to leave this menu.

14. Select [Draw Graph]. When you are finished viewing the graph, press the Esc key to return to the menu. Was the graph drawn on the interval $[0, 4]$? We can adjust the interval over which the functions are graphed, and we can add labels and headings.

15. Select [Change graph specifications] and then from the list of functions select the first one.

16. Select [Minimum x-coordinate on axis] and enter `0`. Note that the default minimum x-value to graph was less than 0. Because you have asked that 0 be the minimum x-coordinate on your graph, Problem Solver automatically sets the minimum x-value to graph equal to 0.

17. Select [Maximum x-coordinate on axis] and enter `4`. This change causes the maximum x-value to graph to be changed as well. This is because the maximum x-value to graph was larger than 4 and you have just requested that the x-axis stop at $x = 4$.

18. Select [Minimum y-coordinate on axis] and enter `-1`. Then select [Maximum y-coordinate on axis] and enter `9`.

19. Select [Label for x-axis] and enter `x`.

20. Select [Label for y-axis] and enter `y`.

21. Press the Esc key to return to the graph functions menu.

22. Select Draw graph. Your graph should be a line passing through the given points.

Exercise Set 2.1

1. Use Problem Solver to graph the equation $y = 2x - 1$ for $x \in [-3, 3]$. Use **x** as a label for the x-axis and **y** as a label for the y-axis.

2. The line with equation $y = -2x + 5$ contains the points (0,5), (1,3), and (2,1). Graph the line and the three points on the same coordinate system.

3. The graph of $y = x^2 - 3x + 2$ contains the points (0,2), (1,0), and (3,2). Graph the equation and the three points on the same coordinate system.

4. Graph each function on the interval $[-1, 1]$:

   (a) $f(x) = 2x^2 - 1$

   (b) $g(x) = 4x^3 - 3x$

   (c) $h(x) = 8x^4 - 8x^2 + 1$

5. Graph each function on the interval $[-\pi, \pi]$:

   (a) $f(x) = x - \frac{1}{6}x^3 + \frac{1}{120}x^5 - \frac{1}{5040}x^7$

   (b) $g(x) = 1 - \frac{1}{2}x^2 + \frac{1}{24}x^4 - \frac{1}{720}x^6$

   You can enter `pi` into Problem Solver for π.

6. Graph the equation $y = x(x-1)(x-2)(x-3)$ for $x \in [-1, 4]$.

7. The line with equation $y = mx + b$ has slope m. Use Problem Solver to graph the equations $y = -3x + 1$, $y = -1x + 1$, $y = 0x + 1$, $y = 1x + 1$, and $y = 3x + 1$, on the same coordinate system. Make some comments concerning the steepness of the graphs and the slopes of the lines.

8. Graph each of the following on the same coordinate system, and then state any observations that you think are appropriate: $y = x^2$, $y = x^2+2$, $y = x^2-1$, and $y = -(x^2 - 1)$. In your comments, compare the graphs of the first and second equations, the first and third equations, and the third and fourth equations.

9. Repeat problem 8 for the following functions, except this time compare all of the graphs with the first graph: $y = x^3$, $y = (x-1)^3$, $y = (x+2)^3$, and $y = (x+2)^3 - 1$.

⋆ 10. By experimenting with the coefficients m and b, find a function of the form $f(x) = mx + b$ whose graph passes through (or as near as you can get) the points $(0, 2)$ and $(4, 6)$. Repeat the problem for the points $(0, 5)$ and $(4, 2)$. Did you make any useful observations that might make another similar problem easier?

⋆⋆ 11. By experimenting with graphs, find values of a, b, and c so that the graph of $f(x) = ax^2 + bx + c$ passes through (or as near as you can get) the points (0,0), (1,2), and (3,2).

12. Graph each of the following functions on the same coordinate system:

    (a) $f(x) = x$
    (b) $g(x) = x^2$
    (c) $h(x) = x^3$
    (d) $u(x) = x^4$
    (e) $v(x) = x^5$

13. Graph each of the following functions:

    (a) $f(x) = x^2 + 1$
    (b) $g(x) = x^3 - x$
    (c) $h(x) = x^4 - x^3 + 2x - 1$

14. Graph a variety of polynomial functions and use these graphs to help you discuss the following question: Is there any relationship between the degree of a polynomial and the number of hills and valleys in its graph?

2.2 Solving Equations

Let $f(x) = x^3 - 2x^2 - x + 1$, and consider the problem of solving the equation $f(x) = 0$ for x. Any x-value where the graph crosses the x-axis is a solution to this equation. From the graph in Figure 2.2, you can see that a solution exists between -2 and 0, a second solution exists between 0 and 2, and a third exists between 2 and 4. First consider the solution between 0 and 2.

Note that $f(0) = 1$ and $f(2) = -1$. The fact that these functional values differ in sign is important. Because of this sign change we know that unless there is a break in the graph of f, the graph must actually pass through the x-axis somewhere between 0 and 2. Since this function is a polynomial, the graph has no breaks. This insures the existence of an x-intercept between 0 and 2. This x-intercept is the solution we seek.

Not knowing the exact location of the solution, we use the midpoint of the interval as an improved estimate. This midpoint is $(0+2)/2$ or 1. Since $f(1) = -1$,

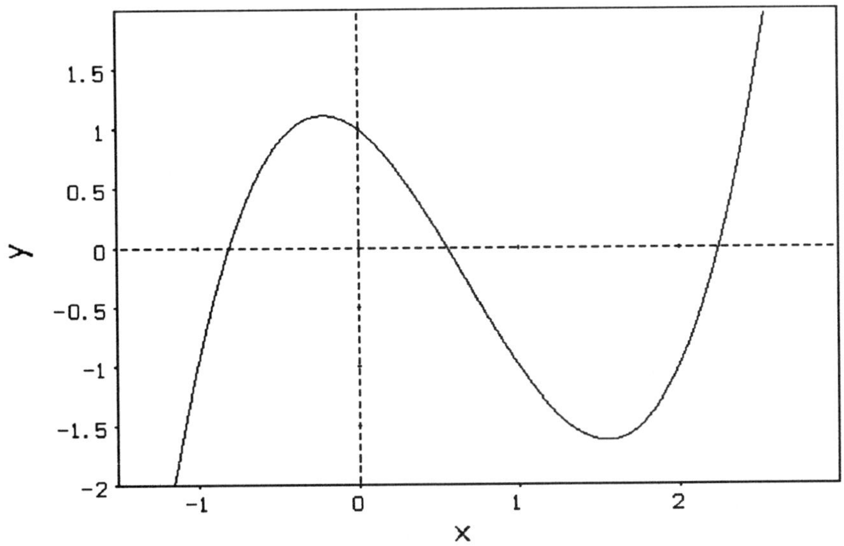

Figure 2.2: Graph of $y = x^3 - 2x^2 - x + 1$

it follows that $f(x)$ changes sign between $x = 0$ and $x = 1$. Consequently the solution we want must be between 0 and 1.

We again refine our estimate by bisecting the new interval [0,1]. This process is repeated until we reach our desired level of accuracy. See Table 2.1 for more details. This process is called the *bisection method.*

2.2.1 Problem Solver Example

We shall solve the example equation $x^3 - 2x^2 - x + 1 = 0$ for $x \in [0, 2]$ using the following steps:

1. Start Problem Solver and select Functions from the main menu.

2. Then select Edit/enter function.

3. Select the first function type `x^3 - 2*x^2 - x + 1` and then press the Enter key. Press the Esc key to return to the functions menu.

4. Now select Solve $f(x) = 0$.

5. From the list of functions, select the function that you entered.

6. You may adjust the maximum acceptable errors if you like. The residual error is $|f(\hat{x})|$, where $\hat{x}$ is the approximate solution. If $\hat{x}$ were the actual solution then the residual error would be zero. The error is $|\hat{x} - x^*|$ where x^* is the actual solution. Since x^* is not known, an upper bound or worst case estimate is given for the error.

Iteration	Interval		Midpoint m	$f(m)$
1	0.000000	2.000000	1.000000	−1.000000
2	0.000000	1.000000	0.500000	0.125000
3	0.500000	1.000000	0.750000	−0.453125
4	0.500000	0.750000	0.625000	−0.162109
5	0.500000	0.625000	0.562500	−0.017334
6	0.500000	0.562500	0.531250	0.054230
7	0.531250	0.562500	0.546875	0.018536
8	0.546875	0.562500	0.554688	0.000621
9	0.554688	0.562500	0.558594	−0.008351
10	0.554688	0.558594	0.556641	−0.003864

Table 2.1: Ten Steps of the Bisection Method

7. Select [Method] and then [Bisection method].
8. Select [Initial guesses] and enter 0 and 2.
9. Finally select [Compute solution].

If you want a paper copy of the results, use the following steps:

1. Select [Options].
2. From the options menu select [Output Device].
3. Then select [Printer].
4. Press the Esc to leave the options menu.
5. Again select [Compute solution]. This time the results will be printed.

Problem Solver should produce the following output:

```
Equation:                         x^3 - 2*x^2 - x + 1 = 0
Algorithm used:                      The bisection method
Initial guess:                                        0 2
Number of iterations:                                  19
Solution:                               0.554958343505859
Error bound:                           3.814697265625e-06
Estimated relative error:              6.873844334921e-06
Residual error:                        4.853950756889e-07
```

Proceeding in a similar manner for the other two solutions, we have:

```
Equation:                        x^3 - 2*x^2 - x + 1 = 0
Algorithm used:                     The bisection method
Initial guess:                                      -2 0
Number of iterations:                                 19
Solution:                              -0.801937103271484
Error bound:                           3.814697265625e-06
Estimated relative error:              4.756853436589e-06
Residual error:                        0.616828786550e-06
```

```
Equation:                        x^3 - 2*x^2 - x + 1 = 0
Algorithm used:                     The bisection method
Initial guess:                                       2 4
Number of iterations:                                 16
Solution:                               2.24697875976562
Error bound:                            3.05175781250e-05
Estimated relative error:               1.35816050741e-05
Residual error:                         4.35380374419e-06
```

2.2.2 Problem Solver Example

We shall use Problem Solver to estimate $\sqrt{5}$ to 10 decimal places of accuracy. Since $\sqrt{5}$ is one of the solutions of the equation $x^2 - 5 = 0$, we can use Problem Solver's equation solver.

1. Start Problem Solver and select [Functions].

2. Next select [Enter/edit function] and select the first function.

3. Enter `x^2 - 5` and press the Enter key.

4. Press the Esc key to return to the functions menu.

5. Select [Solve $f(x) = 0$].

6. From the list of functions in the next menu, select the one that you just entered.

7. Select [Method] and then select [Bisection method].

8. Select [Initial guess] and enter 2 and 3. With $f(x) = x^2 - 5$, $f(2) = -1$, and $f(3) = 4$. Since $f(x)$ changes from negative to positive when x changes from 2 to 3, there must be a point somewhere between 2 and 3 where $f(x) = 0$. This confirms the fact that 2 and 3 were reasonable initial guesses.

9. Now adjust the error tolerances. Select [Maximum error], and enter `0.5e-10`. This will give 10 decimal places of accuracy.

10. Now select [Compute solution].

Problem Solver should display the following information:

```
Equation:                           x^2 - 5 = 0
Algorithm used:            The bisection method
Initial guess:                              2 3
Number of iterations:                        35
Solution:                      2.23606797747198
Error bound:                  2.91038304567e-11
Estimated relative error:     1.30156286615e-11
Residual error:               1.24388499501e-10
```

Rounding to 10 decimal places, $\sqrt{5} \doteq 2.2360679775$.

Exercise Set 2.2

1. Use Problem Solver to solve the equation $x^2 - 9 = 0$.

2. Use Problem Solver to solve the linear equation $3x + 4 = 8$.

3. Find two solutions of the equation $x^2 + x - 3 = 0$.

4. Solve the equation $x^3 - 9 = 0$.

5. Approximate $\sqrt[5]{2}$ using Problem Solver's equation solver. Experiment with the error tolerances and get as much accuracy as you can.

6. Find all solutions of the following equations:

   (a) $x^3 + x^2 + x + 1 = 0$

   (b) $x^4 + x - 1 = 0$

   (c) $x^2 + 1 = 0$

   (d) $x^3 - x = 0$

   (e) $x^4 - x^3 + 2x - 1 = 0$

   (f) $x - \frac{1}{6}x^3 + \frac{1}{120}x^5 - \frac{1}{5040}x^7 = 0$

   (g) $1 - \frac{1}{2}x^2 + \frac{1}{24}x^4 - \frac{1}{720}x^6 = 0$

7. Find the points where the line with equation $x+2y = 4$ intersects the parabola with equation $x^2 - x - 1 = y$.

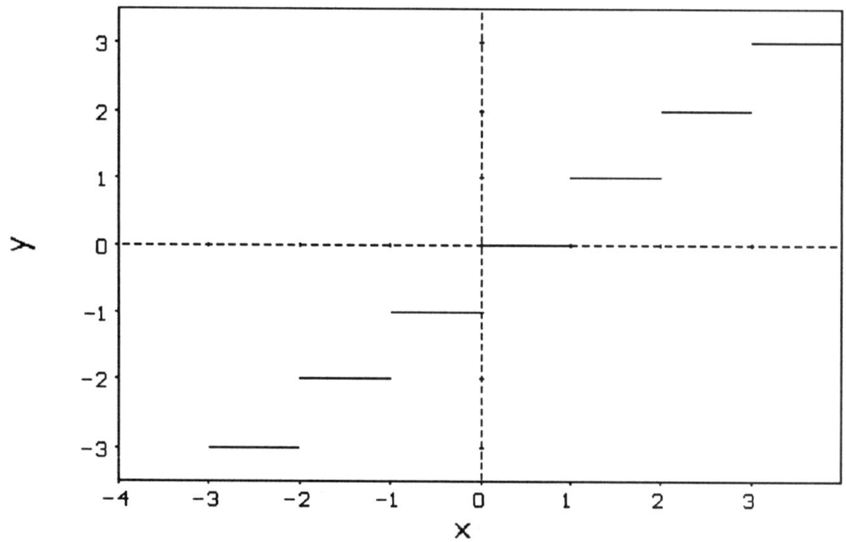

Figure 2.3: The Greatest Integer Function

8. Use Problem Solver to find the points where the circle with equation $x^2 + y^2 = 16$ intersects the graph of $6 = 2x + 3y$.

9. Find all solutions of the equation $8x^4 - 8x^2 + 1 = 0$.

10. Find the points where the graphs of the equations $y = x + 1$ and $y = 1 - x^2$ intersect.

11. Solve the equation $(1 + x)^3 = x$ for x.

12. Wilbur has a rectangular piece of cardboard measuring 24 by 36 inches. He intends to cut a small square from each corner to form four flaps, one on each side. He will then fold the flaps upward to form an open box. If he wants the volume of the resulting box to be 1120 cubic inches, how much should he cut from each corner?

2.3 Applications

The greatest integer function assigns to each x the largest integer not greater than x, and it is denoted by $[\![x]\!]$. For example, $[\![2.5]\!] = 2$, $[\![4]\!] = 4$, and $[\![-4.6]\!] = -5$. The graph of the greatest integer function has the stair-step shape shown in Figure 2.3.

2.3.1 Problem Solver Example

We shall use Problem Solver to graph the greatest integer function and to generate a table of functional values for $x \in \{-1.0, -0.9, -0.8, \ldots, 1\}$.

1. Start Problem Solver and select Functions.

2. From the functions menu select Enter/edit function.

3. Select the first function and type `gint(x)`.

4. Press the Esc key to return to the functions menu.

5. Select Graph functions and then select Function selection. Select the function that you just entered and press `Y` to indicate that you wish to graph that function.

6. Press the Esc key to return to the graph functions menu.

7. Select Draw graph. Notice that the stair steps are incorrectly connected with vertical lines. When you have finished viewing the graph, press the Esc key to return to the graph functions menu.

8. Select Change graph specifications and then select the function that you entered earlier.

9. Select Connect points with line segments and enter `N` to indicate that the points are not to be connected. Press the Esc key to leave this menu.

10. Again select Draw graph. This time the graph should be correct. Whenever the correct graph has breaks, it is not appropriate to have Problem Solver connect the points it draws with line segments.

11. Press the Esc key twice to return to the functions menu.

12. From the functions menu select Functional values.

13. From the list of functions select the function that you entered.

14. Select Number of values. Since there are 21 numbers in the set {-1.0,-0.9,...,1.0}, enter `21`.

15. Select Enter x-values from the keyboard.

16. Enter the x-values in the first column of the matrix editor.

17. Press the Esc key when you are through.

18. Select Calculate functional values. You should see a table of values. The first column consists of the x-values, and the second column consists of the corresponding $f(x)$ values.

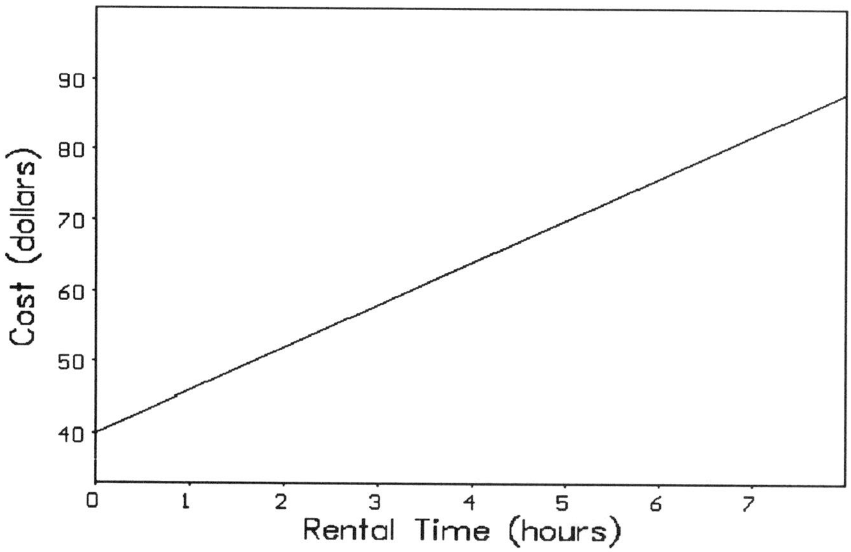

Figure 2.4: Charlie's First Charging Policy

Hours of Use	Cost
$0 \leq x < 1$	$40 + 6$
$1 \leq x < 2$	$40 + 6(2)$
$2 \leq x < 3$	$40 + 6(3)$

Table 2.2: Charlie's Second Charging Policy

2.3.2 The Chain Saw

Charlie's U-rent has chain saws that can be rented for a nonrefundable deposit of \$40 plus \$6 per hour for up to 8 hours. We shall express the cost of the chain saw as a function of the time that it is rented and graph this cost function.

Let y represent the cost of renting the chain saw for x hours. Since the total cost must include both the \$40 deposit and \$6 for each of the x hours of use, we have $y = 6x + 40$ for $0 \leq x \leq 8$. The relationship between x and y is an example of a linear function. The equation $y = 6x + 40$ is a mathematical model of Charlie's charging policy for the chain saw, and its graph is shown in Figure 2.4.

At a recent national convention of the U-rent Society, Charlie learned that it would be both easier and more profitable if he charged the \$40 deposit and then \$6 for each hour or any portion of an hour for up to 8 hours. He liked the idea of making more money for less work, and he immediately adopted this new plan. His new charging policy for the first three hours is shown in Table 2.2. Using the greatest integer function to model this new policy, we have

$$y = 40 + 6[\![x + 1]\!] \text{ for } 0 \leq x \leq 8.$$

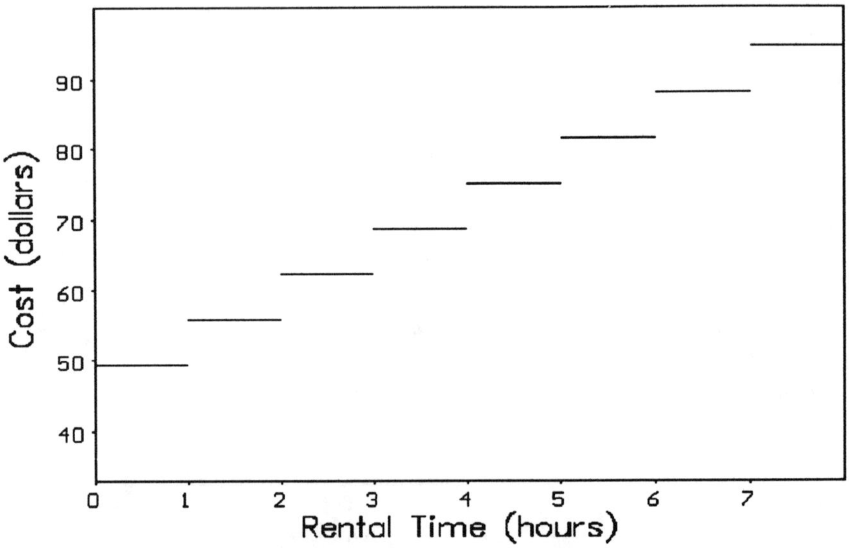

Figure 2.5: Charlie's Final Charging Policy

Finally, suppose that Charlie must collect state tax on all rentals at a rate of 7.3% of the rental cost. To incorporate the tax into our model, we note that the tax on $40 + 6[\![x+1]\!]$ dollars is $0.073(40 + 6[\![x+1]\!])$ dollars. Adding this tax to the last expression for y, we arrive at our final version of the chain saw rental model. It is given by the equation

$$y = 40 + 6[\![x+1]\!] + 0.073(40 + 6[\![x+1]\!])$$

or

$$y = 1.073(40 + 6[\![x+1]\!]).$$

The graph of Charlie's final charging policy is shown in Figure 2.5.

2.3.3 The Flying Ball

Slick Rock enjoys his reputation as the most successful pitcher ever for the Blue Mountain Weasels. While relaxing one day, Slick threw a baseball straight up into the air with an initial velocity of 96 ft/sec. If he released the ball at a height of 7 feet, then the function

$$s(t) = -16t^2 + 100t + 7$$

gives the height of the ball t seconds after it is thrown. This model ignores the effects of air resistance and assumes that once the ball is in the air the only force acting on it is gravity. The graph of the height of the ball at time t is given in Figure 2.6. To find how long the ball is in the air, we solve the equation $s(t) = 0$. From the graph we can see that the answer is between 6 and 8 seconds. In fact,

Figure 2.6: Height of the Ball as a Function of Time

$s(6) > 0$ and $s(8) < 0$. This means that we can use the bisection method with 6 and 8 as initial guesses. Doing this yields the following output:

```
Equation:                     -16*t^2 + 100*t + 7 = 0
Algorithm used:                  The bisection method
Initial guess:                                    6 8
Number of iterations:                              30
Solution:                             6.3192330840975
Error bound:                        1.862645149231e-09
Estimated relative error:           2.947581018840e-10
Residual error:                     7.131097845559e-08
```

The ball is in the air for about 6.3 seconds.

2.3.4 The Phaser GT

Wilbur loves fine automobiles, and he often tells the story of his first car. One day he found an advertisement in the Mill Creek Chronicle for a good used Phaser GT owned by a woman named Maria Garcia. Wil was sure that this car would soon become a collector's item, so he rushed to the owner's house to have a look. Unfortunately, the price of this car was \$3500, and Wil had only \$3100. Although Wil usually leaves out this detail, his job history and credit rating put a standard automobile loan out of the question. The same two items also made potential cosigners wary. Consequently Wil had to make a cash deal.

After an extended discussion, Maria agreed to hold the Phaser GT for two years while Wil raised additional money. In addition, she agreed to hold the price at

\$3500. This was a great deal for Maria who considered the car junk, and it seemed like a good deal for Wil who was sure that the car would soon start appreciating in value.

As Wil was thinking about finding a job, he struck upon a brilliant idea. Why not put his \$3100 into a savings account and let his money grow. This would work if his \$3100 would grow to \$3500 in two years.

Wil remembered the compound interest equation $A = P(1+i)^n$, where P is the principal (deposit), i is the interest rate per period, n is the number of periods and A is the total amount of both the principal and the interest in the account after n periods.

For his situation the deposit would be $P = \$3100$. At that time banks compounded interest monthly, so $n = (12)(2) = 24$. The i in the formula represents the percent interest rate per period expressed as a number between 0 and 1. That is, i is the percent interest rate divided by 100. Since the compounding was done monthly, i represents the monthly interest rate as a percent divided by 100.

Wil decided to express this equation in terms of the more familiar annual interest rate r as a percent. Since i is the monthly rate, he set

$$i = \frac{r}{(12)(100)}$$

Wil's new equation was

$$A = 3100\left(1 + \frac{r}{(12)(100)}\right)^{24}$$

where r represents the annual rate of interest expressed as a percent.

Wil studied the relationship between A and r by producing the graph shown in Figure 2.7. He wondered what interest rate would be required for him to have \$3500 after 24 months. He needed to determine r so that $A = 3500$.

Setting $A = 3500$ produces the equation

$$3100\left(1 + \frac{r}{(12)(100)}\right)^{24} - 3500 = 0$$

After entering the left side of this equation into Problem Solver we find the interest rate r that solves this equation.

```
Equation:                    3100*(1 + 0.01*r/12)^24 - 3500 = 0
Algorithm used:                           The bisection method
Initial guess:                                             4 8
Number of iterations:                                       30
Solution:                                       6.083410885185
Error bound:                                3.725290298462e-09
Estimated relative error:                   6.123686807896e-10
Residual error:                             9.175801096717e-08
```

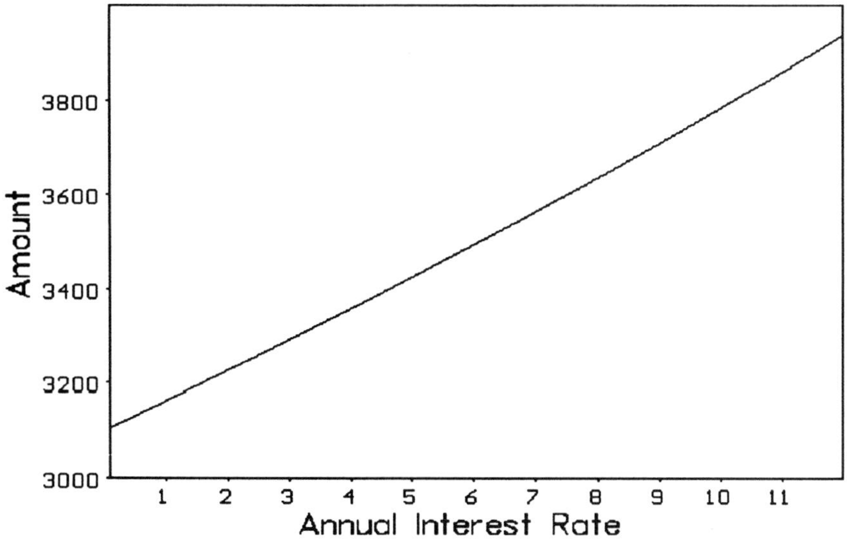

Figure 2.7: Amount A as a Function of r

Fortunately, one bank paid 6.1% annual interest, and Wilbur's plan worked. He still has his Phaser GT, and it is as shiny as it was the day he bought it. He expects it to begin soon increasing in value.

Exercise Set 2.3

1. Use Problem Solver to solve the equation $x^3 - 5x + 1 = 0$.

2. Use Problem Solver to construct a table of values for the function $f(x) = [\![x^2 - 3x]\!] + 1$ for $x \in \{1.0, 1.1, 1.2, \ldots, 3.0\}$.

3. Wilbur rode a taxi from his house to Maria Garcia's house to look at her Phaser GT automobile. The charges were \$2.00 plus \$1.50 for each eighth mile. Find a linear function that gives the total cost for riding this taxi for x miles. As an example of the use of your function, calculate the cost of riding the 5.5 miles from Wil's house to Maria's house. Use Problem Solver to graph this function for $0 \leq x \leq 20$.

4. Maria is selling tickets for a trip to Hawaii for a group of 50-100 tourists. The cost per ticket is \$800 minus a discount of \$5 for each ticket sold in excess of 50.

   (a) Find a function $R(n)$ that gives the total revenue to Maria if n tickets are sold.

   (b) Graph the function $R(n)$.

(c) Create a table of values for $R(n)$ for $n \in \{50, 55, 60, \ldots, 100\}$.

5. The equation $F = \frac{9}{5}C + 32$ relates the temperature using the Fahrenheit scale to the temperature using the Celsius scale.

   (a) Graph this relationship for $-50 \leq C \leq 150$.

   (b) Create a table of values for $0 \leq C \leq 100$ in steps of five degrees.

6. One day Slick Rock was trying to sell his crop of potatoes at a roadside vegetable stand. He charged ten cents a pound for each whole pound and gave the customer any fraction of a pound for free. Use the greatest integer function to model Slick's charging policy for the potatoes.

7. Repeat the last problem with this modification: Now suppose that Slick rounds to the nearest pound and charges 10 cents per pound on this rounded weight. Again use the greatest integer function to model his charging policy.

8. See Example 2.3.4. Wilbur made an impulse purchase when he decided to buy a \$1500 computer system to help him manage his new savings account. He held back \$1500 from his planned deposit of \$3100, and he purchased the computer with it. He place the remaining \$1600 into his savings account.

   (a) Without additional income how far short of his needed \$3500 would he be in two years?

   (b) Wilbur was offered a job selling potatoes at a roadside vegetable stand. This meant that he could make monthly payments into his savings account. If he put P dollars into his account each month, these deposits (not counting the original \$1600 deposit and the interest it would earn) would grow to a value of

   $$V = P\frac{(1+i)^n - 1}{i}$$

   after n months where i represents the monthly interest rate expressed as a number between 0 and 1 rather than as a percent. Convert i to r as in Example 2.3.4, and find the monthly payment required in order that the payments plus the interest over the two-year period will make up for his computer purchase.

   (c) Use Problem Solver to draw a graph of the value V at the end of the two years as a function of the payment P. The vertical axis could be labeled "Future Value" and the horizontal axis "Monthly Payment." Select an appropriate heading and subheading to give this graph an attractive appearance.

9. A certain rifle has a muzzle velocity of 2300 ft/sec. If this gun is fired straight up into the air, and if the muzzle is 9 feet from the ground, then the equation giving the height of the bullet t seconds after it is fired is

$$y = -16t^2 + 2300t + 9$$

assuming once again that all forces other than gravity can be neglected. How long will the bullet be in the air? Is your answer realistic? Discuss.

2.4 Multiline Functions

Sometimes it is awkward to define a function on a single line. For example consider the function

$$f(x) = 12 \text{ if } x \leq 3, \text{ and } f(x) = 4x \text{ if } x > 3$$

We can rewrite the above line more clearly as follows:

$$f(x) = \begin{cases} 12 & \text{if } x \leq 3 \\ 4x & \text{if } x > 3 \end{cases}$$

Notice that $f(2) = 12$ since $2 \leq 3$, and $f(6) = (4)(6) = 24$ since $6 > 3$.

2.4.1 Problem Solver Example

We illustrate the use of Problem Solver for graphing the function given above for $x \in [-3, 6]$.

1. Start Problem Solver and select Functions.
2. Select Enter/edit function.
3. Select the first function and enter `12`.
4. Select the second function, select Formula, and then enter `4*x`. Press the Esc key to return to the functions menu.
5. Select Graph functions.
6. Select Change graph specifications. Then select your first function.
7. Select Minimum x-value to graph and enter `-3`.
8. Select Maximum x-value to graph and enter `0`.
9. Select Minimum x-coordinate on axis and enter `-3`.

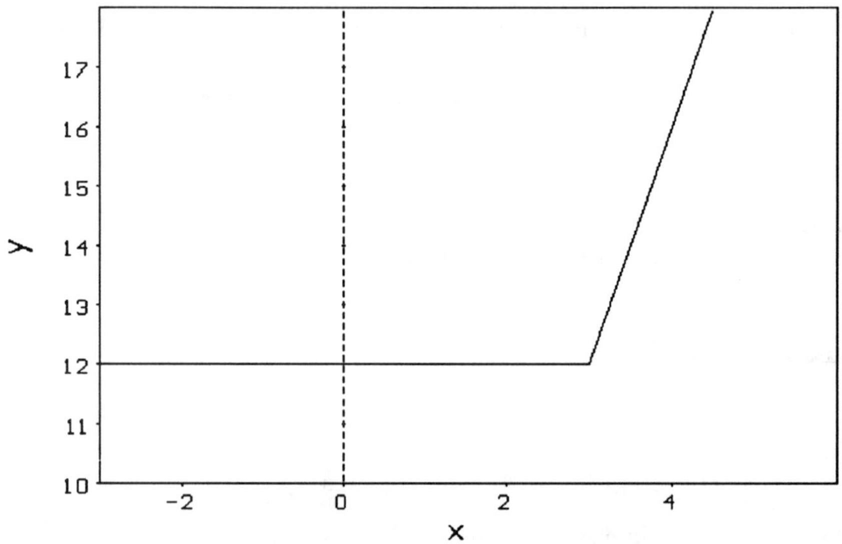

Figure 2.8: Multiline Function

10. Select [Maximum x-coordinate on axis] and enter 6.

11. Select [Minimum y-coordinate on axis] and enter 10.

12. Select [Maximum y-coordinate on axis] and enter 18.

13. Press the Esc key to leave this menu.

14. We must now adjust the second function. Again select [Change Graph Specifications]. This time select the second function.

15. Select [Minimum x-value to graph] and enter 3.

16. Select [Maximum x-value to graph] and enter 6.

17. Press the Esc key to leave this menu.

18. Select [Function selection]. Select the two functions that you entered.

19. Leave this menu by pressing the Esc key.

20. Select [Draw Graph]. Your graph should look like Figure 2.8.

There is a way to enter this function into a single line. The Problem Solver function $step(x)$ is defined as follows:

$$step(x) = \begin{cases} 0 & \text{if } x < 0 \\ 1 & \text{if } x \geq 0 \end{cases}$$

Using this function, we could write

$$f(x) = 12\,step(3 - x) + 4x(1 - step(3 - x)).$$

This function would be entered into Problem Solver as follows:

```
12*step(3-x)+4*x*(1-step(3-x))
```

The next example illustrates the use of a multiline function.

2.4.2 The Mystery Writer

Melanie has just completed a deal with her publisher for her latest novel *Murder at Bone Springs*. Her publisher has agreed to give her \$1000 cash plus 3% of all sales on the first 5000 copies and 7% on all sales above 5000. They plan to market this potential best-seller as a paperback for \$4.95.

To represent Melanie's income as a function of the number of books sold, we let n represent the number of books sold and $I(n)$ represent the corresponding income for Melanie. Then

$$I(n) = \begin{cases} 1000 + 0.03(4.95)n & \text{if } n \leq 5000 \\ 1000 + 0.03(4.95)(5000) + 0.07(4.95)(n - 5000) & \text{if } n > 5000 \end{cases}$$

or more simply

$$I(n) = \begin{cases} 1000 + 0.1485n & \text{if } n \leq 5000 \\ 1742.50 + 0.3465(n - 5000) & \text{if } n > 5000 \end{cases}$$

To calculate Melanie's income from the first 3000 books, we note that $3000 < 5000$ and use the first line of the definition. Her income is

$$I(3000) = 1000 + (0.1485)(3000) = 1445.50.$$

To calculate her income from the first 8000 books we use the second line of the definition since $8000 > 5000$. This amount is

$$I(8000) = 1742.50 + 0.3465(8000 - 5000) = 2782.00.$$

Exercise Set 2.4

1. Graph the function f defined by

$$f(x) = \begin{cases} x - 1 & \text{if } x \leq 2 \\ x & \text{if } x > 2 \end{cases}$$

2. Graph the function f defined by

$$f(x) = \begin{cases} x^2 - 1 & \text{if } x \leq 2 \\ 4x - 5 & \text{if } x > 2 \end{cases}$$

3. Graph the function f defined by

$$f(x) = \begin{cases} x^2 - 1 & \text{if } x \leq 1 \\ x & \text{if } x > 1 \end{cases}$$

4. During the summer months, Mr. Greene operates a lawn service. He will send a worker to your house for \$15 per hour for the first two hours and \$8 for each additional hour. Use a multiline function to model Mr. Greene's charging system. Use Problem Solver to graph the cost as a function of time for 0 to 8 hours.

5. A bank sells certificates of deposit at various interest rates depending on how much is spent on the certificate. For \$1000 to \$5000 certificates the interest rate is 7.8%, and for certificates costing more than \$5000 but less than \$10000 the interest rate is 8.8%. For any certificate of at least \$10000, the interest rate is 10%.

   (a) Find a multiline function $f(x)$ that gives the amount in the bank after 1 year as a function of x where x represents the cost of the certificate.

   (b) Graph this function for $1000 \leq x \leq 15000$.

6. Maria is selling tickets for a trip to Hawaii for a group of up to 100 tourists. The cost per ticket is \$800 minus a discount of \$5 for each ticket sold in excess of 50.

   (a) Find a multiline function $R(n)$ that gives the total revenue to Maria if n tickets are sold.

   (b) Graph the function $R(n)$.

Chapter 3

Matrices

3.1 Solving Systems of Equations

An $m \times n$ matrix is a rectangular array of numbers consisting of m rows and n columns. For example,

$$\begin{pmatrix} 5 & 3 & 1 \\ 4 & 0 & 2 \end{pmatrix}$$

is a 2×3 matrix. As our first application of matrices we shall consider systems of linear equations. The system

$$\begin{array}{rcrcrcr} 4x & + & y & & & = & 0 \\ 2x & + & 3y & - & z & = & -2 \\ 8x & - & 4y & + & z & = & 13 \end{array}$$

can be represented more efficiently by the *augmented matrix*

$$\begin{pmatrix} 4 & 1 & 0 & 0 \\ 2 & 3 & -1 & -2 \\ 8 & -4 & 1 & 13 \end{pmatrix}.$$

3.1.1 Problem Solver Example

We shall use Problem Solver to solve the system

$$\begin{array}{rcrcl} 2x & - & y & = & 3 \\ 4x & + & y & = & 9. \end{array}$$

Use the following steps to solve this problem with Problem Solver:

1. Start Problem Solver and select Matrices from the main menu.

2. Select [Linear equations/inverses].

3. From the next menu select [Edit/enter matrix]. This matrix has 2 rows and 3 columns. Enter these numbers as they are requested.

4. You should now be viewing the matrix editor. Enter numbers into the cells by typing the number and pressing the Enter key. Move to new cells by using the arrow keys located on the right of your keyboard. After you enter the augmented matrix, the screen should look like the display below:

	1	2	3
1	2	−1	3
2	4	1	9

Now press the Esc key to leave the matrix editor.

5. Select [Compute solution] from the menu and [Gaussian elimination] from the next menu. The solution $x = 2$ and $y = 1$ will be displayed in a matrix. The first number displayed is x, the first variable, and the second number displayed is y, the second variable. If you want to print your results, use the following steps.

6. Select [Options] and then from the options menu select [Output device]. From the next list of choices select [Printer].

7. Press the Esc key to leave the options menu.

8. Repeat step 5.

3.1.2 Problem Solver Example

We shall use Problem Solver to solve the system

$$\begin{array}{rcrcl} 2x & - & y & = & 3 \\ 4x & - & 2y & = & 9. \end{array}$$

Use the following steps to solve this problem with Problem Solver:

1. Start Problem Solver and select [Matrices] from the main menu. item Select [Linear equations/inverses].

2. From the next menu select [Edit/enter matrix]. This matrix has 2 rows and 3 columns. Enter these numbers as they are requested.

3. You are now in the matrix editor. After you enter the augmented matrix, the screen should look as follows:

	1	2	3
1	2	−1	3
2	4	−2	9

Now press the Esc key to leave the matrix editor.

4. Select [Compute solution] from the menu and [Gaussian elimination] from the next menu.

This system has no solution. The error message indicates that the coefficient matrix is a singular matrix. This means that either the system has multiple solutions, or it has none at all. Singular matrices will be defined in the next section. To find out whether or not the system has solutions, select [Compute solution] and then [Reduced-row echelon form] from the next menu. Problem Solver will give you the reduced matrix, and then you can examine the system represented by the reduced matrix to see whether or not it has solutions. If any of the rows consist of all zero entries except for the last column, and if the last column entry in that row is nonzero, then that row represents an equation that equates 0 to a nonzero number. Since that is impossible, there are no solutions to such systems. This example is such a system.

One of the methods used by Problem Solver is called *Gaussian elimination.* This method involves the use of the following elementary row operations on matrices:

1. Interchange any two rows.
2. Multiply the elements of any row by any nonzero number.
3. Replace any row with the sum of itself and any other row.

Operations 2 and 3 are often combined into one step:

> Replace any row with the sum of itself and any multiple of any other row.

Our objective is to perform elementary row operations on the augmented matrix until the elements below the diagonal are all zero. We now trace through the steps of applying Gaussian elimination to the example given at the beginning of this section. The original augmented matrix is

$$\begin{pmatrix} 4.000 & 1.000 & 0.000 & 3.000 \\ 2.000 & 3.000 & -1.000 & -2.000 \\ 8.000 & -4.000 & 1.000 & 13.000 \end{pmatrix}.$$

We first reduce the elements below the diagonal in the first column to 0. In order to better control rounding errors, Problem Solver uses a method called *partial pivoting.* Rows are interchanged so that the element in column 1 with largest magnitude is moved into the row 1 position. Thus rows 1 and 3 are interchanged, and the new matrix is

$$\begin{pmatrix} 8.000 & -4.000 & 1.000 & 13.000 \\ 2.000 & 3.000 & -1.000 & -2.000 \\ 4.000 & 1.000 & 0.000 & 3.000 \end{pmatrix}.$$

We are now ready to create zeros below the diagonal in column 1. We do this by replacing row 2 with the sum of row 2 and $-2/8$ times row 1, and row 3 with the sum of row 3 and $-4/8$ times row 1. Our new matrix is

$$\begin{pmatrix} 8.000 & -4.000 & 1.000 & 13.000 \\ 0.000 & 4.000 & -1.250 & -5.250 \\ 0.000 & 3.000 & -0.500 & -3.500 \end{pmatrix}.$$

We next turn our attention to column 2. At this stage we no longer consider row 1 or column 1. We are now looking only at the lower right hand 2×3 matrix. Row 2 is used to reduce the element(s) in column 2 below the diagonal to 0. Since 4 is no smaller that 3 we need not interchange rows 2 and 3. To reduce the 3 in column 2 to 0 we replace row 3 with the sum of row 3 and $-3/4$ times row 2. This produces the matrix

$$\begin{pmatrix} 8.000 & -4.000 & 1.000 & 13.000 \\ 0.000 & 4.000 & -1.250 & -5.250 \\ 0.000 & 0.000 & 0.438 & 0.438 \end{pmatrix}.$$

Our reduction process is complete. This augmented matrix represents the system

$$\begin{array}{rcrcrcr} 8x & - & 4y & + & z & = & 13.000 \\ & & 4y & - & 1.25z & = & -5.250 \\ & & & & 0.438z & = & 0.438. \end{array}$$

This system is easy to solve if we start with the last equation and work our way back toward the first. This process is called *backward substitution.*

From the last equation we have

$$z = 0.438/0.438 = 1.$$

Substituting $z = 1$ into the second equation yields

$$4y - 1.25 = -5.25$$

or

$$4y = -4$$

or

$$y = -1.$$

Substituting $y = -1$ and $z = 1$ into the first equation leads to

$$8x - 4(-1) + 1 = 13$$

or

$$8x = 8$$

and finally

$$x = 1.$$

The solution is

$$x = 1, \; y = -1, \text{ and } z = 1.$$

The Gauss-Jordan method is similar to Gaussian-elimination. Rather than using backward substitution, the Gauss-Jordan method performs row operations on the matrix until all elements not on the diagonal are 0. In addition, the elements on the diagonal are changed to 1 by multiplying the rows by appropriate nonzero numbers.

We illustrate the Gauss-Jordan method with the same example. The original augmented matrix is

$$\begin{pmatrix} 4.000 & 1.000 & 0 & 3.000 \\ 2.000 & 3.000 & -1.000 & -2.000 \\ 8.000 & -4.000 & 1.000 & 13.000 \end{pmatrix}.$$

Interchanging rows 1 and 3 we have

$$\begin{pmatrix} 8.000 & -4.000 & 1.000 & 13.000 \\ 2.000 & 3.000 & -1.000 & -2.000 \\ 4.000 & 1.000 & 0 & 3.000 \end{pmatrix}.$$

After reducing column 1 and multiplying the first row by 1/8, our matrix is

$$\begin{pmatrix} 1.000 & -0.500 & 0.125 & 1.625 \\ 0.000 & 4.000 & -1.250 & -5.250 \\ 0.000 & 3.000 & -0.500 & -3.500 \end{pmatrix}.$$

We next reduce column 2 and multiply row 2 by 1/4 to get

$$\begin{pmatrix} 1.000 & 0.000 & -0.031 & 0.969 \\ 0.000 & 1.000 & -0.312 & -1.312 \\ 0.000 & 0.000 & 0.438 & 0.438 \end{pmatrix}.$$

Finally column 3 is reduced and row 3 is multiplied by 1/0.438. Our last matrix is

$$\begin{pmatrix} 1.000 & 0.000 & 0.000 & 1.000 \\ 0.000 & 1.000 & 0.000 & -1.000 \\ 0.000 & 0.000 & 1.000 & 1.000 \end{pmatrix}.$$

No backward substitution is required; the solution may be read directly from the last column.

3.1.3 Problem Solver Example

Problem Solver allows you to control the row operations in the Gauss-Jordan method. We shall illustrate this using the system solved in the last section.

1. Start Problem Solver and select Matrices.
2. Select Matrix operations.
3. Select Enter/edit matrix and then select the first matrix.
4. This matrix has 3 rows and 4 columns. Enter these numbers as they are requested.
5. You should now be viewing the matrix editor. When you have finished entering the augmented matrix, the screen should look like the display below:

	1	2	3	4
1	4	1	0	3
2	2	3	−1	−2
2	8	−4	1	13

   Press the Esc key to leave the matrix editor.
6. We need to have more decimal places displayed. Select Options and then select Format for text output.
7. Select Decimal places in matrices and enter `6`. Press the Esc key twice to return to the matrix operations menu.
8. Select User controlled matrix reduction and then select the first matrix.
9. To control round-off error, rows 1 and 3 should be interchanged. Select First row index and enter 1. Select Second row index and enter 3. Select Interchange rows 1 and 3.
10. Select Multiplier and enter `1/8`.
11. Select Row 1 ← 0.125*Row1.
12. Select Multiplier and enter `-4`.
13. Select Row 3 ← Row 3 + −4*Row 1.
14. Select Second row index and enter `2`.

15. Select [Multiplier] and enter `-2`.

16. Select [Row 2 ← Row 2 + −2*Row 1]. Column 1 should now be reduced. It should have 1 in the first position and 0 in all other positions. We now start reducing column 2.

17. Select [First row index] and enter `2`.

18. Select [Second row index] and enter `3`.

19. Select [Multiplier] and enter `1/4`.

20. Select [Row 2 ← 0.25*Row 2].

21. Select [Multiplier] and enter `-3`.

22. Select [Row 3 ← Row 3 + −3*Row 2].

23. Select [Second row index] and enter 1.

24. Select [Multiplier] and enter `1/2`.

25. Select [Row 1 ← Row 1 + 0.5*Row 2]. The second column should now be reduced. The second element should be 1, and the first and third should be 0.

26. Select [First row index] and enter `3`.

27. Select [Multiplier] and enter `1/.4375`.

28. Select [Row 3 ← 2.28571*Row 3].

29. Select [Multiplier] and enter `.03125`.

30. Select [Row 1 ← Row 1 + 0.03125*Row 3].

31. Select [Second row index] and enter 2.

32. Select [Multiplier] and enter `.3125`.

33. Select [Row 2 ← Row 2 + 0.3125*Row 3]. The third column should now be reduced. It should have 1 in the last position and 0 in the first two position. The solution $x = 1$, $y = -1$, and $z = 1$ can be read from the last column.

3.1.4 A Curve Fitting Problem

We shall find the equation of the parabola that passes through the three points (0,1), (2,0), and (6,8). Our equation will have the form

$$y = ax^2 + bx + c$$

for some choice of a, b, and c. We must choose these values so that the graph of the resulting equation will pass through the three given points.

Since (0,1) is on the graph, we know that the above equation will be a true statement when $x = 0$ and $y = 1$. Replacing x with 0 and y with 1 yields

$$0a + 0b + c = 1.$$

Using the points (2,0) and (6,8) in a similar manner we get the two additional equations

$$\begin{array}{rcrcrcl} 2^2a & + & 2b & + & c & = & 0 \\ 6^2a & + & 6b & + & c & = & 8. \end{array}$$

The augmented matrix for these three equations is

$$\begin{pmatrix} 0 & 0 & 1 & 1 \\ 4 & 2 & 1 & 0 \\ 36 & 6 & 1 & 8 \end{pmatrix}.$$

Solving this system with Problem Solver, we get the solution

$$a = 0.417, \ b = -1.333, \text{ and } c = 1.$$

Thus the parabola has equation

$$y = 0.417x^2 - 1.333x + 1.$$

The graph of this polynomial along with the three given points is shown in Figure 3.1.

The polynomial found in this example is called an *interpolating* polynomial. Given the positive integer n and the $n+1$ points $(x_0, y_0), (x_1, y_1), \ldots, (x_n, y_n)$, where the x_is are distinct, it is possible to find a polynomial of degree at most n whose graph passes through the given points. The $n + 1$ coefficients could be determined by solving a system of $n + 1$ equations.

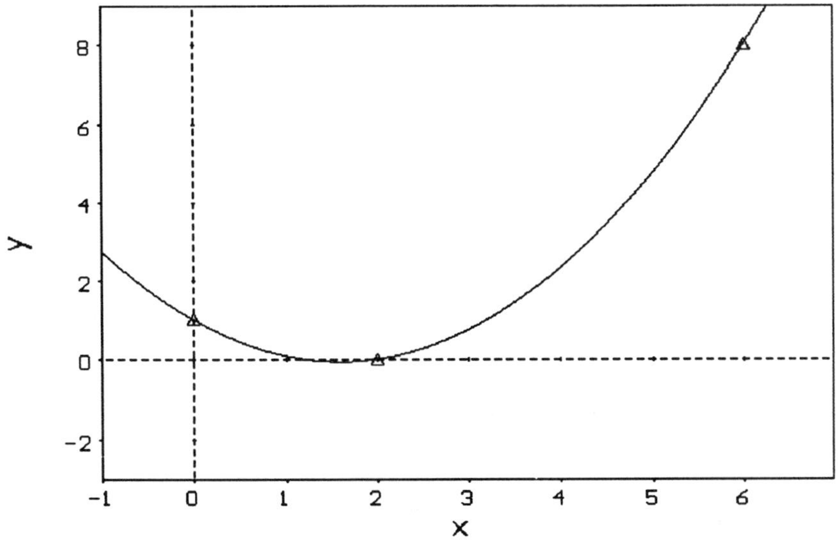

Figure 3.1: Polynomial Fit for Points

Exercise Set 3.1

1. Use Problem Solver to solve the first example given in this section.

2. Solve the system

$$\begin{array}{rcrcrcr} 3x & - & 4y & + & 6z & = & 0 \\ 4x & - & 2y & + & 5z & = & 2 \\ x & + & 3y & - & 4z & = & 3 \end{array}$$

using Problem Solver and Gaussian elimination.

⋆ 3. Consider the following system:

$$\begin{array}{rcrcrcr} x & + & 2y & + & 3z & = & 4 \\ 5x & + & 6y & + & 7z & = & 8 \\ 9x & + & 10y & + & 11z & = & 12 \end{array}$$

(a) Use Problem Solver and the Gauss-Jordan method to solve this problem.

(b) Solve this problem again and this use Problem Solver to do the arithmetic, but control the row operations yourself. See Problem Solver Example 3.1.3.

4. Solve the system

$$\begin{array}{rcrcrcrcr} w & + & x & + & y & + & z & = & 0 \\ w & - & x & - & y & + & z & = & 2 \\ w & + & x & + & 3y & - & 4z & = & 3 \\ 3w & + & x & - & y & + & z & = & 5 \end{array}$$

using Problem Solver and Gaussian elimination.

5. Solve the system

$$\begin{array}{rcrcrcrcr} 2w & + & 3x & - & 4y & + & 6z & = & 0 \\ 5w & - & 4x & - & 2y & + & 5z & = & 2 \\ w & + & x & + & 3y & - & 4z & = & 3 \\ w & + & 2x & - & 3y & + & 8z & = & 10 \end{array}$$

using Problem Solver and the Gauss-Jordan method.

6. Use a system of equations to find the equation of the line that passes through the points (2,4) and (5,6).

7. Find the equation of the second degree polynomial that passes through the points $(-2,-1)$, (2,7), and (3,14).

8. Find the equation of the third degree polynomial that passes through the points (1,1), (2,2), (3,1), and (4,2).

★ 9. Three sisters are to divide an inheritance of $300,000 according to the following two rules:

- The third sister is to receive 10% less than the average of what the first two sisters receive.
- The first sister is to pay an inheritance tax of $50,000. After paying the $50,000 tax she is still to have 10% more than the second sister.

Use a system of equations to find how much each sister is to receive.

10. Mr. Greene assigns two types of problems to his physics students: easy and moderately hard. The easy problems usually require about 5 minutes for a student to solve while the moderately hard problems require 15 minutes. Mr. Greene likes to assign 10 problems per day, and he expects his students to work for 90 minutes on each assignment. How many of each type of problem should he assign per day?

11. Jake has just won $100,000 playing the Oregon Lotto game, and now he is considering three investments. One investment is very safe, but its return is only 6.2%; a second investment is fairly safe, and its return is 8.5%; and a third investment, although somewhat risky, returns 15%. Jake has decided to invest all of his money in these accounts in such a way that his overall return will be 8%. In addition, he wants to put 80% of his money in the two safer investments. How much money should Jake put into each of the three investments?

★ 12. During the last spring thaw a bridge over the Grand River collapsed, and it

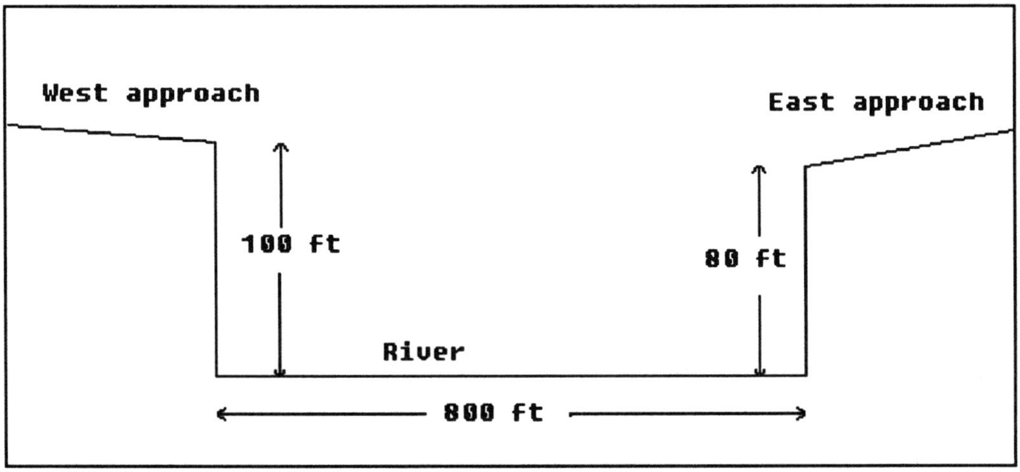

Figure 3.2: Side View of River and Roadway

is your job to find an appropriate shape for the road surface of a replacement bridge.

On the west side of the river, the road slopes down toward the river at an angle of 5 degrees from the horizontal, and the broken edge of the road surface is 100 meters above the water level. On the east side of the river, the road slopes upward away from the river at an angle of 7 degrees above the horizontal, and the broken edge of the road surface is 80 meters above the water level. The river is 800 meters wide. See Figure 3.2.

(a) Place the origin of a coordinate system at the water level on the west side of the river, use kilometers as units, and find the equation of a line that joins the two approaches. This equation will be used to build the road surface on the bridge.

(b) Draw the graph of your line and the two approaches.

(c) Discussion question: Do you think that this line will make a satisfactory shape for the road surface? Think about a car approaching this bridge at high speed.

We shall be looking at this problem again. As we develop further techniques, it will be possible to improve our design for the road surface, especially at the points where the new road joins the old road.

3.2 Matrix Operations

The sum of two $m \times n$ matrices A and B is the $m \times n$ matrix formed by adding corresponding components of A and B.

3.2.1 Problem Solver Example

Find the sum of the matrices

$$A = \begin{pmatrix} 2 & 3 & 4 \\ -3 & 8 & 6 \end{pmatrix} \text{ and } B = \begin{pmatrix} 5 & 8 & 8 \\ -7 & 2 & 4 \end{pmatrix}.$$

1. Start Problem Solver and select [Matrices] from the main menu.
2. Select [Matrix operations] from the next menu.
3. Select [Enter/edit matrix].
4. Select [A].
5. You will be prompted to enter the number of rows and columns of A. This matrix has 2 rows and 3 columns. Enter these numbers as they are requested.
6. Now enter the matrix A using the arrow keys on the right side of the keyboard to move the cursor to various positions in the matrix. When you are finished, the edit screen should look like following display:

	1	2	3
1	2	3	4
3	−3	8	6

   Press the Esc key to leave the editor.
7. Again select [Enter/edit matrix]. Enter the matrix B in exactly the same way you entered A.
8. Finally select [SUM ← A+B] to calculate the sum of A and B.

The $n \times n$ identity matrix is the $n \times n$ matrix with 1 in each diagonal position and 0 elsewhere. For example,

$$\begin{pmatrix} 1 & 0 & 0 \\ 0 & 1 & 0 \\ 0 & 0 & 1 \end{pmatrix}$$

is the 3×3 identity matrix.

If A is an $n \times n$ matrix then a matrix B satisfying $AB = BA = I$ is called the *inverse* of A and is denoted by A^{-1}. Matrices that have no inverse are called *singular matrices*.

3.2.2 Problem Solver Example

Let

$$A = \begin{pmatrix} 2 & 3 & 0 \\ 7 & 4 & 2 \\ 5 & 0 & 1 \end{pmatrix}, \text{ and let } B = \begin{pmatrix} 3 \\ -3 \\ 5 \end{pmatrix}.$$

Find the matrix product AB and A^{-1}, the inverse of the matrix A.

1. Start Problem Solver and select [Matrices] from the main menu.
2. Select [Matrix operations].
3. Select [Enter/edit matrix].
4. Select [A].
5. You will be prompted to enter the number of rows and columns of A. This matrix has 3 rows and 3 columns. Enter these numbers as they are requested.
6. Now enter the matrix A using the arrow keys on the right side of the keyboard to move the cursor to various positions in the matrix. When you are finished, the edit screen should look like this:

	1	2	3
1	2	3	0
2	7	4	2
3	5	0	1

   Press the Esc key to leave the editor.

7. Again select [Enter/edit matrix] to enter the matrix B.
8. Select B from the next menu. The matrix B has 3 rows and 1 column. Enter these numbers as they are requested.
9. Now enter the matrix B. After you are finished, the edit screen should look like this:

	1
1	3
2	−3
3	5

   Press the Esc key to leave the matrix editor.

10. Select [PROD ← A*B] to calculate the product of A and B.

11. To find the inverse of A select A ← Inverse of A. This will replace A with A^{-1}.

12. Print matrices from this menu by selecting Print matrix.

3.2.3 Grade Calculations

When Mr. Greene taught physics at Mill Creek High School in the late 1960s, he formed his final grades from the following categories with the indicated weights:

Category	Weight
Homework	.15
Laboratories	.35
Tests	.45
Attendance	.05

He had 7 students with the following records:

Name	Homework	Laboratories	Tests	Attendance
Bobby Jo	89	95	92	99
Calvin	85	95	80	75
Millie	86	90	75	80
Norbert	95	99	92	100
Olaf	75	89	81	90
Tabatha	89	90	99	95
Winnefred	76	78	82	80

In order to impress his students with the utility of mathematics, Mr. Greene used matrix multiplication to calculate their final averages. He defined two matrices as follows:

$$G = \begin{pmatrix} 89 & 95 & 92 & 99 \\ 85 & 95 & 80 & 75 \\ 86 & 90 & 75 & 80 \\ 95 & 99 & 92 & 100 \\ 75 & 89 & 81 & 90 \\ 89 & 90 & 99 & 95 \\ 76 & 78 & 82 & 80 \end{pmatrix} \text{ and } W = \begin{pmatrix} .15 \\ .35 \\ .45 \\ .05 \end{pmatrix}.$$

To calculate the grade averages Mr. Greene found the matrix product

$$GW = \begin{pmatrix} 92.95 \\ 85.75 \\ 82.15 \\ 95.30 \\ 83.35 \\ 94.15 \\ 79.60 \end{pmatrix}.$$

The final averages are listed in order with Bobby Jo's 92.95 listed first and Winnefred's 79.6 listed last.

3.2.4 Matrix Codes

At a recent Mill Creek High School reunion, some of Mr. Greene's former students shared fond memories of the use of matrices on one of Mr. Greene's physics tests.

It all started when Mr. Greene showed the students how to use matrices to encode and decode messages. He first assigned a number to each letter of the alphabet. The letter A was assigned 1, the letter B was assigned 2, and so on until Z was assigned the number 26. Using this code, the message "Please put the cat out" becomes "16 12 5 1 19 5 16 21 20 20 8 5 3 1 20 15 21 20."

This code, like all codes where each letter is assigned a different number, has a serious flaw. Letters occur in the English language with known frequencies. This means that if a long enough encoded message is intercepted, statistical analysis can be used to break the code.

What is needed is a code where 1 might correspond to A sometimes and 2 might correspond to A at other times. Mr. Greene used matrices to accomplish this.

He first set

$$A = \begin{pmatrix} 2 & 0 & 1 \\ 3 & 1 & 1 \\ 0 & 1 & 1 \end{pmatrix}.$$

Next he separated the numerical codes for "Please put the cat out" into groups of three numbers by defining the following 3×1 matrices:

$$\begin{pmatrix} 16 \\ 12 \\ 5 \end{pmatrix} \quad \begin{pmatrix} 1 \\ 19 \\ 5 \end{pmatrix} \quad \begin{pmatrix} 16 \\ 21 \\ 20 \end{pmatrix} \quad \begin{pmatrix} 20 \\ 8 \\ 5 \end{pmatrix} \quad \begin{pmatrix} 3 \\ 1 \\ 20 \end{pmatrix} \quad \begin{pmatrix} 15 \\ 21 \\ 20 \end{pmatrix}.$$

Had the number of letters in the message not been a multiple of three, Mr. Greene would have padded the end of the message with enough Zs to get a multiple of three. One by one he multiplied each of these matrices by A:

$$A\begin{pmatrix} 16 \\ 12 \\ 5 \end{pmatrix} = \begin{pmatrix} 37 \\ 65 \\ 17 \end{pmatrix}, \text{ and } A\begin{pmatrix} 1 \\ 19 \\ 5 \end{pmatrix} = \begin{pmatrix} 7 \\ 27 \\ 24 \end{pmatrix}.$$

Continuing in this manner with each of the 3×1 matrices he finally had the encoded message "37 65 17 7 27 24 52 89 41 45 73 13 26 30 21 50 86 41."

"This code," he asserted, "is nearly impossible to break." The students couldn't help but agree. After all, the t in *put* was encoded as 17 but the t immediately following in *the* was encoded as 7.

Several students were puzzled about how to decode the message. Mr. Greene explained that the inverse of A could be used for decoding in exactly the same way that A was used for encoding. First he found the inverse of A to be

$$A^{-1} = \begin{pmatrix} 0 & \frac{1}{3} & -\frac{1}{3} \\ -1 & \frac{2}{3} & \frac{1}{3} \\ 1 & -\frac{2}{3} & \frac{2}{3} \end{pmatrix}.$$

Then he divided the encoded message into 3×1 matrices

$$\begin{pmatrix} 37 \\ 65 \\ 17 \end{pmatrix} \begin{pmatrix} 7 \\ 27 \\ 24 \end{pmatrix} \begin{pmatrix} 52 \\ 89 \\ 41 \end{pmatrix} \begin{pmatrix} 45 \\ 73 \\ 13 \end{pmatrix} \begin{pmatrix} 26 \\ 30 \\ 21 \end{pmatrix} \begin{pmatrix} 50 \\ 86 \\ 41 \end{pmatrix}.$$

By multiplying A^{-1} by each of these matrices, Mr. Greene recovered the original message. He found the products

$$A^{-1}\begin{pmatrix} 37 \\ 65 \\ 17 \end{pmatrix} = \begin{pmatrix} 16 \\ 12 \\ 5 \end{pmatrix} \text{ and } A^{-1}\begin{pmatrix} 7 \\ 27 \\ 24 \end{pmatrix} = \begin{pmatrix} 1 \\ 19 \\ 5 \end{pmatrix}$$

and continued in this way until the entire message was decoded.

Later that week when Mr. Greene administered a chapter test on the properties of sound, his astonished students found that the entire test was encoded. Their fear quickly subsided as they noticed that the kindly Mr. Greene had put the encoding matrix A on the blackboard!

3.2.5 The Mill Creek Burger Wars

When Wilbur was a high school student, he helped to celebrate the grand openings of his town's only two fast food restaurants. Burgers 'n Things and Bob's Cheap Burgers each gained 250 loyal customers that day. Each year after the openings, Bob lost half of his customers to Burgers 'n Things, and Burgers 'n Things lost one third of its customers to Bob. Wilbur couldn't help but wonder if eventually Burgers 'n Things would have all 500 customers. To study this problem, Wilbur constructed the following matrix:

	Bob's	Burgers 'n Things
Bob's	$\frac{1}{2}$	$\frac{1}{2}$
Burgers 'n Things	$\frac{1}{3}$	$\frac{2}{3}$

Each entry represents the fraction of customers who move from the row restaurant to the row restaurant. Another matrix was used to give the customer count for each restaurant:

$$\begin{array}{cc} \text{Bob's} & \text{Burgers 'n Things} \\ \end{array}$$

$$\begin{pmatrix} 250 & 250 \end{pmatrix}$$

After 1 year Bob's Cheap Burgers would have half of their original customers plus a third of Burgers 'n Things' customers. Thus Bob's Cheap Burgers would have

$$\frac{1}{2}250 + \frac{1}{3}250 = 208.333333$$

customers. Similarly Burgers 'n Things would have

$$\frac{1}{2}250 + \frac{2}{3}250 = 291.666667$$

customers.

These values can be calculated using matrix multiplication. After one year the customer count would be

$$\begin{pmatrix} 250 & 250 \end{pmatrix} \begin{pmatrix} \frac{1}{2} & \frac{1}{2} \\ \frac{1}{3} & \frac{2}{3} \end{pmatrix} = \begin{pmatrix} 208.333333 & 291.666667 \end{pmatrix}$$

After 2 years the customer count would be

$$\begin{pmatrix} 208.333333 & 291.666667 \end{pmatrix} \begin{pmatrix} \frac{1}{2} & \frac{1}{2} \\ \frac{1}{3} & \frac{2}{3} \end{pmatrix} = \begin{pmatrix} 201.388888 & 298.611111 \end{pmatrix}$$

After 3 years the customer count would be

$$\begin{pmatrix} 201.388888 & 298.611111 \end{pmatrix} \begin{pmatrix} \frac{1}{2} & \frac{1}{2} \\ \frac{1}{3} & \frac{2}{3} \end{pmatrix} = \begin{pmatrix} 200.231481 & 299.768518 \end{pmatrix}$$

After 4 years the customer count would be

$$\begin{pmatrix} 200.231481 & 299.768518 \end{pmatrix} \begin{pmatrix} \frac{1}{2} & \frac{1}{2} \\ \frac{1}{3} & \frac{2}{3} \end{pmatrix} = \begin{pmatrix} 200.038580 & 299.961420 \end{pmatrix}$$

It appears that in the long run Burgers 'n Things will have 300 customers and Bob will have to settle for 200.

3.2.6 Problem Solver Example

Problem Solver can be used to easily carry out the computations of the last example. The required steps follow:

1. Start Problem Solver and select [Matrices].
2. Select [Matrix operations].

3. Select [Enter/edit matrix] and then select [A].

4. Matrix A has 2 rows and 2 columns. Enter these numbers as they are requested.

5. Enter the matrix A into the matrix editor. Be sure to enter the numbers as fractions rather than decimals. Then 2/3 and 1/3 will be accurate to 15 decimal places. When you are finished, the edit screen should look like the following display:

	1	2
1	0.500	0.500
2	0.333	0.667

   Press the Esc key to leave the editor.

6. You need to enter the matrix giving the original customer distribution. Again select [Enter/edit matrix]. This time select [PROD].

7. This matrix has 1 row and 2 columns. Enter these numbers as they are requested.

8. Enter the matrix into the matrix editor. When you are finished, the edit screen should look like this:

	1	2
1	250	250

   Press the Esc key to leave the editor.

9. select [PROD ← PROD * A]. The distribution at the end of the first year should be displayed.

10. Repeatedly select [PROD ← PROD * A] to see the customer distribution for the years ahead.

Exercise Set 3.2

1. Let $A = \begin{pmatrix} 2 & 3 & 7 \\ 5 & 1 & 8 \end{pmatrix}$ and $B = \begin{pmatrix} 7 & 6 & 1 \\ -4 & 3 & 2 \end{pmatrix}$. Use Problem Solver to find each of the following:

   (a) $5A$

   (b) $A + B$

(c) $A - B$

(d) $3A - 7B$

2. Let $A = \begin{pmatrix} 2 & 3 \\ 2 & 1 \end{pmatrix}$ and $B = \begin{pmatrix} 3 \\ -4 \end{pmatrix}$. Use Problem Solver to find each of the following:

(a) AB

(b) $A^{-1}B$

3. Let $A = \begin{pmatrix} 1 & \frac{1}{2} & \frac{1}{3} & \frac{1}{4} & \frac{1}{5} \\ \frac{1}{2} & \frac{1}{3} & \frac{1}{4} & \frac{1}{5} & \frac{1}{6} \\ \frac{1}{3} & \frac{1}{4} & \frac{1}{5} & \frac{1}{6} & \frac{1}{7} \\ \frac{1}{4} & \frac{1}{5} & \frac{1}{6} & \frac{1}{7} & \frac{1}{8} \\ \frac{1}{5} & \frac{1}{6} & \frac{1}{7} & \frac{1}{8} & \frac{1}{9} \end{pmatrix}$.

(a) Find A^{-1}.

(b) Find the inverse of A^{-1}.

Suggestion: Remember that fractions can be entered directly into Problem Solver. There is no need to convert the fractions to decimals.

4. The matrix $A = \begin{pmatrix} 2 & 4 & 5 \\ 1 & 1 & 1 \\ 0 & 0 & 2 \end{pmatrix}$ was used to encode the the secret message 105 33 10 103 25 10 131 36 38 158 47 16 91 25 2 234 58 52.

Use Problem Solver to decode this message.

5. The matrix $A = \begin{pmatrix} 2 & 1 & 3 \\ 6 & 2 & 1 \\ 8 & 1 & 4 \end{pmatrix}$ was used to encode the following secret messages:

(a) 63 68 113 13 29 38 56 147 177 132 162 242

(b) 63 62 95 35 79 78 97 174 229 49 55 64 101 137 210

Use Problem Solver to decode these messages.

6. Suppose that when Bob's Cheap Burgers and Burgers 'n Things first opened, Bob's low prices attracted all 500 customers. Again assume that each year Bob loses 1/2 of his customers to Burgers 'n Things and Burgers 'n Things loses 1/3 of its customers to Bob.

(a) What can be said about the long term distribution of customers?

(b) Repeat the problem assuming that Burgers 'n Things got all 500 customers in the beginning.

3.3 Linear Programming

Problem Solver solves linear programming problems of the following form: Maximize (or minimize) the objective function

$$Z = c_1x_1 + c_2x_2 + \cdots + c_nx_n$$

subject to the functional constraints

$$\begin{array}{ccccccc} a_{1,1}x_1 & + & a_{1,2}x_2 & +\cdots+ & a_{1,n}x_n & \leq & b_1 \\ a_{2,1}x_1 & + & a_{2,2}x_2 & +\cdots+ & a_{2,n}x_n & \leq & b_2 \\ \vdots & & & & & & \\ a_{m,1}x_1 & + & a_{m,2}x_2 & +\cdots+ & a_{m,n}x_n & \leq & b_m \end{array}$$

and the nonnegativity constraints

$$x_1 \geq 0, x_2 \geq 0, ..., x_n \geq 0.$$

Any of the *less than or equal to* signs in the functional constraints can be replaced with either *equal to* or *greater than or equal to* signs.

3.3.1 Problem Solver Example

We shall find the maximum of the objective function

$$Z = 2x + 3y$$

subject to the constraints

$$\begin{array}{ccccc} x & + & y & \leq & 5 \\ x & + & 2y & \leq & 8 \end{array}$$

$$x \geq 0, y \geq 0.$$

1. Start Problem Solver and select Matrices from the main menu.

2. From the matrices menu select Linear programming.

3. From the linear programming menu select Enter/edit model.

4. This model has 2 decision variables (x and y) and 2 constraints. The nonnegativity constraints $x \geq 0$ and $y \geq 0$ are not counted. Enter these numbers as they are requested.

5. The screen before you is the model editor. You can enter data into the cells by typing the number or symbol that you want and then pressing the Enter key. Use the arrow keys on the right side of the keyboard to move to a new cell.

   The coefficients of the objective function should be put in the first row. Thus 2 and 3 go on the first row. The coefficients of the constraints go in the other rows. Enter values until the screen looks like the display that follows:

	1	2	3	4
obj	2	3		
1	1	1	<	5
2	1	2	<	8

   The nonnegativity constraints are assumed and need not be entered. After completing the data entry, press the Esc key to leave the model editor.

6. You should again be seeing the linear programming menu on the screen. Select [Compute optimal solution] from this menu.

7. The solution is $x = 2$, $y = 3$, and $Z = 13$.

If you wish to print the solution, use the following steps:

1. Select [Options] from either the linear programming menu or the main menu.

2. Select [Output device].

3. Select [Printer].

4. Press the Esc key to return to the linear programming menu.

5. Select [Compute optimal solution]. This time the solution will be printed.

3.3.2 Problem Solver Example

Consider the problem of finding the minimum of the objective function

$$Z = x + 9y$$

subject to the constraints

$$\begin{array}{rcrcr} -3x & + & y & \geq & 8 \\ 2x & + & 3y & \geq & 10 \\ 4x & + & y & \geq & 5 \end{array}$$

$$x \geq 0, y \geq 0.$$

We shall show the steps required for using Problem Solver to solve this problem and print the results.

1. Start Problem Solver and select Matrices from the main menu.

2. From the matrices menu select Linear programming.

3. From the linear programming menu select Enter/edit model.

4. This model has 2 decision variables (x and y) and 3 constraints. Enter these numbers as they are requested.

5. Enter this problem into the model editor. The coefficients of the objective function should be placed in the first row. The coefficients of the constraints go in the other rows. Enter values until the screen looks like the display below:

	1	2	3	4
obj	1	9		
1	−3	1	>	8
2	2	3	>	10
2	4	1	>	5

   The nonnegativity constraints are assumed and need not be entered. After completing the data entry, press the Esc key to leave the model editor.

6. You should again be seeing the linear programming menu on the screen. Select Options.

7. Then select Output device and from the next menu select Printer.

8. Press the Esc key to return to the linear programming menu.

9. Select Objective. You will be asked whether to minimize the objective function. Respond with `Y`.

10. Select Compute optimal solution.

11. Problem Solver gives the solution $x = 0$, $y = 8$, and $Z = 72$.

	Small	Medium	Large
Plywood (sheets)	1.5	2.5	5.0
Assembly (man-hours)	2.5	3.5	4.5
Finish (man-hours)	4.0	4.5	5.5

Table 3.1: Requirements Table for Doghouses

3.3.3 Ace Manufacturing

When Wilbur needed to purchase a doghouse, he immediately thought of Ace Manufacturing. They had an excellent reputation for their line of high-quality products and could never keep enough doghouses in stock. Wilbur visited Ace one day, and as usual they were out of medium size doghouses. When he asked the salesman why they didn't build more, the salesman explained the method used to decide how many of each of the three sizes to build.

The profit for the small, medium, and large models was \$12, \$15, and \$19, respectively. The plywood requirements for the models were 1.5 sheets, 2.5 sheets, and 5 sheets. One problem for Ace was brought on by a lumber strike. Only 20 sheets of plywood per day were available.

The salesman went on to state that each model required a certain amount of time in assembly and some additional time in the finish and paint room. The requirements are given in Table 3.1. The assembly room could devote up to 15 man-hours per day to the doghouses, and the finish room had up to 20 man-hours per day. The salesman went on to explain to Wilbur that Ace likes to maximize their profit. He was now writing hurriedly on the back of a credit application form as Wilbur tried to follow along.

The salesman defined three decision variables as follows: Let x_1 represent the number of small doghouses to produce per day, let x_2 represent the number of medium doghouses to produce per day, and let x_3 represent the number of large doghouses to produce per day.

If the profit is represented by Z, then $Z = 12x_1 + 15x_2 + 19x_3$. Ace wanted to find values of the decision variables that would maximize Z subject to the following constraints:

Plywood constraint:	$1.5x_1 + 2.5x_2 + 5.0x_3 \leq 20$
Assembly room constraint:	$2.5x_1 + 3.5x_2 + 4.5x_3 \leq 15$
Finish room constraint:	$4.0x_1 + 4.5x_2 + 5.5x_3 \leq 20$
Nonnegativity constraints:	$x_1 \geq 0$ and $x_2 \geq 0$

Using Problem Solver, we find that the maximum value of Z is 65.88, and the optimal values of the decision variables are $x_1 = 1.765$, $x_2 = 0$, and $x_3 = 2.353$. This means that the maximum daily profit from the doghouses was \$65.88 and that

	Oregon warehouse	California warehouse
Factory 1	10	15
Factory 2	12	17
Factory 3	14	20

Table 3.2: Shipping Costs for Ace Manufacturing

Ace could realize this profit by building small doghouses at a rate of 1.765 per day and large doghouses at a rate of 2.533 per day.

3.3.4 A Transportation Problem

Several years after Wilbur looked for a doghouse at Ace Manufacturing, Ace got serious about selling doghouses. They set up three factories in western Washington and two warehouses: one in Oregon and one in California. The maximum daily output of the three factories was 600, 700, and 500 units. The daily demand for the warehouses was 600 and 1000 for the Oregon and California warehouses, respectively. The shipping costs per unit are given in Table 3.2. We shall determine how many units should be shipped from each of the three factories to each of the two warehouses if we wish to meet all of the above requirements and at the same time minimize the total shipping cost. To solve this problem, we define the following decision variables:

Variable	Meaning
x_1	Number of units to ship from Factory 1 to the Oregon warehouse.
x_2	Number of units to ship from Factory 2 to the Oregon warehouse.
x_3	Number of units to ship from Factory 3 to the Oregon warehouse.
x_4	Number of units to ship from Factory 1 to the California warehouse.
x_5	Number of units to ship from Factory 2 to the California warehouse.
x_6	Number of units to ship from Factory 3 to the California warehouse.

The total shipping cost per day is

$$Z = 10x_1 + 12x_2 + 14x_3 + 15x_4 + 17x_5 + 20x_6.$$

Since Factory 1 can produce at most 600 units, no more than 600 units can be shipped from Factory 1. Thus

$$x_1 + x_4 \leq 600.$$

Applying similar reasoning to Factories 2 and 3 yields the constraints

$$x_2 + x_5 \leq 700$$

and

$$x_3 + x_6 \leq 500.$$

Since the Oregon warehouse requires 600 units per day,

$$x_1 + x_2 + x_3 \geq 600.$$

Since the California warehouse requires 1000 units per day,

$$x_4 + x_5 + x_6 \geq 1000.$$

The transportation problem is modeled by the following linear programming problem: Find the minimum value of the function

$$Z = 10x_1 + 12x_2 + 14x_3 + 15x_4 + 17x_5 + 20x_6$$

subject to the constraints

$$x_1 + x_4 \leq 600$$

$$x_2 + x_5 \leq 700$$

$$x_3 + x_6 \leq 500$$

$$x_1 + x_2 + x_3 \geq 600$$

$$x_4 + x_5 + x_6 \geq 1000$$

$$x_i \geq 0, \text{ for } i = 1, 2, \ldots, 6.$$

Using Problem Solver, we find that the solution is $Z = 23,600$, $x_1 = 0$, $x_2 = 300$, $x_3 = 300$, $x_4 = 600$, $x_5 = 400$, and $x_6 = 0$. This means that the minimum daily shipping cost is \$23,600, and in order to have the minimum shipping cost, Ace should ship nothing from Factory 1 to the Oregon warehouse, 300 units per day from Factory 2 to the Oregon warehouse, 300 units per day from Factory 3 to the Oregon warehouse, 600 units per day from Factory 1 to the California warehouse, 400 units per day from Factory 2 to the California warehouse, and nothing from Factory 3 to the California warehouse.

Factories 1 and 2 are working at full capacity, but Factory 3 is producing only 300 units per day rather than its maximum of 500 units per day. Notice that what is best for the organization may not be best for Factory 3.

The transportation problem is such an important application of linear programming that special methods have been developed to handle it. For problems with many variables, these special techniques are far more efficient than the standard linear programming approach.

Item	Price
Hamburger	.79
Chicken bits	1.25
French fries	.98
Vanilla shake	1.49
Orange juice	.78
Sausage biscuit	1.89
Scrambled eggs	.98

Table 3.3: Burgers 'n Things' Menu

Item	Vitamin A	Vitamin C	Calcium	Iron
Hamburger	2.0	3.0	8.4	15.8
Chicken bits	0.0	3.5	1.1	7.0
French fries	0.0	20.9	0.9	3.4
Vanilla shake	7.0	0.0	32.0	4.7
Orange juice	0.0	120.0	0.0	0.0
Sausage biscuit	8.4	0.0	11.9	19.1
Scrambled eggs	13.0	2.0	6.1	14.1

Table 3.4: Recommended Daily Allowance (%) of Vitamins, Calcium, and Iron

Item	Cal	Cholesterol (mg)	Fat (g)	Protein (g)
Hamburger	263	29.1	11.3	12.4
Chicken bits	323	62.5	20.2	19.1
French fries	220	8.57	11.5	3.0
Vanilla shake	352	30.6	8.4	9.3
Orange juice	80	0.0	0.0	1.3
Sausage biscuit	585	285.3	39.9	19.8
Scrambled eggs	180	205.0	13.0	13.2

Table 3.5: Calories, Cholesterol, Fat, and Protein at Burgers 'n Things

Item	Minimum	Maximum
Calories	1200	2000
Cholesterol (mg)	0	300
Fat (g)	0	50
Protein (g)	20	50

Table 3.6: Dietary Constraints

3.3.5 Slick Rock's Diet

Slick Rock likes fast food. He is especially fond of several items available at Burgers 'n Things. These are listed in Table 3.3 with their prices. The numbers give the percentage of the recommended daily allowance of the various nutrients in each item. In spite of the fact that Slick has been taking all of his meals at Burgers 'n Things, he considers himself to be quite serious about health. Following his doctor's recommendation he has decided to impose some limitations on his diet. These are shown in Table 3.6. In addition, he wants to get between 95% and 150% of the recommended daily allowance for vitamins A and C, calcium, and iron.

Being concerned about expenses, Slick wants to find a selection of food items that meets all of the above requirements and at the same time minimizes the cost. He first defined the decision variables:

Variable	Meaning
x_1	number of burgers per day
x_2	number of chicken bit servings per day
x_3	number of French fry servings per day
x_4	number of vanilla shakes per day
x_5	number of orange juice servings per day
x_6	number of sausage biscuits per day
x_7	number of servings of scrambled eggs per day

The daily cost of this diet is

$$Z = 0.79x_1 + 1.25x_2 + 0.98x_3 + 1.49x_4 + 0.78x_5 + 1.89x_6 + 0.98x_7$$

The restrictions on the diet can be expressed as inequalities.
Calorie limit:

$$263x_1 + 323x_2 + 220x_3 + 352x_4 + 80x_5 + 585x_6 + 180x_7 \leq 2000$$

$$263x_1 + 323x_2 + 220x_3 + 352x_4 + 80x_5 + 585x_6 + 180x_7 \geq 1200$$

Cholesterol limit:

$$29.1x_1 + 62.5x_2 + 8.57x_3 + 30.6x_4 + 0x_5 + 285.3x_6 + 205x_7 \leq 300$$

Fat limit:

$$11.3x_1 + 20.2x_2 + 11.5x_3 + 8.4x_4 + 0x_5 + 39.9x_6 + 13x_7 \leq 50$$

Protein requirements:

$$12.4x_1 + 19.1x_2 + 3x_3 + 9.3x_4 + 1.3x_5 + 19.8x_6 + 13.2x_7 \geq 20$$

$$12.4x_1 + 19.1x_2 + 3x_3 + 9.3x_4 + 1.3x_5 + 19.8x_6 + 13.2x_7 \leq 50$$

Vitamin A requirements:

$$2x_1 + 0x_2 + 0x_3 + 7x_4 + 0x_5 + 8.4x_6 + 13x_7 \geq 95$$

$$2x_1 + 0x_2 + 0x_3 + 7x_4 + 0x_5 + 8.4x_6 + 13x_7 \leq 150$$

Vitamin C requirements:

$$3x_1 + 3.5x_2 + 20.9x_3 + 0x_4 + 120x_5 + 0x_6 + 2x_7 \geq 95$$

$$3x_1 + 3.5x_2 + 20.9x_3 + 0x_4 + 120x_5 + 0x_6 + 2x_7 \geq 150$$

Calcium requirements:

$$8.4x_1 + 1.1x_2 + 0.9x_3 + 32x_4 + 0x_5 + 11.9x_6 + 6.1x_7 \geq 95$$

$$8.4x_1 + 1.1x_2 + 0.9x_3 + 32x_4 + 0x_5 + 11.9x_6 + 6.1x_7 \leq 150$$

Iron requirements:

$$15.8x_1 + 7x_2 + 3.4x_3 + 4.7x_4 + 0x_5 + 19.1x_6 + 14.1x_7 \geq 95$$

$$15.8x_1 + 7x_2 + 3.4x_3 + 4.7x_4 + 0x_5 + 19.1x_6 + 14.1x_7 \leq 150$$

In addition to the above constraints each decision variable must be nonnegative.

Running this problem through Problem Solver reveals that there are no feasible solutions. This means that it is impossible to meet all of the dietary requirements if one eats only Slick Rock's favorite items at Burgers 'n Things. Being a sensible person, Slick Rock modified his diet and his requirements. He decided to:

1. eat only one meal per day at Burgers 'n Things,
2. add a garden salad and a chef's salad to his diet,
3. remove the scrambled eggs from his diet, and
4. replace the vanilla shakes with skimmed milk.

Item	Minimum	Maximum
Calories	200	800
Cholesterol (mg)	0	150
Fat (g)	0	30
Protein (g)	8	20

Table 3.7: New Dietary Constraints

Item	Vitamin A	Vitamin C	Calcium	Iron
Milk	10	4	30	0
Garden salad	15	15	10	4
Chef's salad	25	35	20	8

Table 3.8: Recommended Daily Allowance (%) in the New Items

Since he would be eating only once per day at Burgers'n Things, he modified his requirements table as shown in Table 3.7. In addition he decided on a minimum of 15% and a maximum of 50% for his recommended daily allowance of vitamins A and C, calcium, and iron. The modified variable definitions are as follows:

Variable	Meaning
x_1	number of burgers per day
x_2	number of chicken bit servings per day
x_3	number of French fry servings per day
x_4	number of servings of milk
x_5	number of orange juice servings per day
x_6	number of sausage biscuits per day
x_7	number of servings of the garden salad per day
x_8	number of servings of the chef's salad per day

Item	Cal	Cholesterol (mg)	Fat (g)	Protein (g)
Milk	90	5	0.6	8.8
Garden salad	91	110	5.5	6.1
Chef's salad	273	125	13.0	21.0

Table 3.9: Cholesterol, Fat, and Protein in New Items

Item	Price
Milk	.80
Garden salad	1.50
Chef's salad	2.50

Table 3.10: Prices for New Items

The nutritional and price information that he needs for the three new items is given in Tables 3.8, 3.9, and 3.10. Thus the new linear programming problem is: Minimize

$$Z = 0.79x_1 + 1.25x_2 + 0.98x_3 + 0.80x_4 + 0.78x_5 + 1.89x_6 + 1.5x_7 + 2.5x_8$$

subject to the constraints:

$$\begin{aligned}
263x_1 + 323x_2 + 220x_3 + 90x_4 + 80x_5 + 585x_6 + 91x_7 + 273x_8 &\leq 800 \\
263x_1 + 323x_2 + 220x_3 + 90x_4 + 80x_5 + 585x_6 + 91x_7 + 273x_8 &\geq 200 \\
29.1x_1 + 62.5x_2 + 8.6x_3 + 5x_4 + 0x_5 + 285.3x_6 + 110x_7 + 125x_8 &\leq 150 \\
11.3x_1 + 20.2x_2 + 11.5x_3 + 0.6x_4 + 0x_5 + 39.9x_6 + 5.5x_7 + 13x_8 &\leq 30 \\
12.4x_1 + 19.1x_2 + 3x_3 + 8.8x_4 + 1.3x_5 + 19.8x_6 + 6.1x_7 + 21x_8 &\geq 8 \\
12.4x_1 + 19.1x_2 + 3x_3 + 8.8x_4 + 1.3x_5 + 19.8x_6 + 6.1x_7 + 21x_8 &\leq 20 \\
2x_1 + 0x_2 + 0x_3 + 10x_4 + 0x_5 + 8.4x_6 + 15x_7 + 25x_8 &\geq 15 \\
2x_1 + 0x_2 + 0x_3 + 10x_4 + 0x_5 + 8.4x_6 + 15x_7 + 25x_8 &\leq 50 \\
3x_1 + 3.5x_2 + 20.9x_3 + 4x_4 + 120x_5 + 0x_6 + 15x_7 + 35x_8 &\geq 15 \\
3x_1 + 3.5x_2 + 20.9x_3 + 4x_4 + 120x_5 + 0x_6 + 15x_7 + 35x_8 &\leq 50 \\
8.4x_1 + 1.1x_2 + 0.9x_3 + 30x_4 + 0x_5 + 11.9x_6 + 10x_7 + 20x_8 &\geq 15 \\
8.4x_1 + 1.1x_2 + 0.9x_3 + 30.0x_4 + 0x_5 + 11.9x_6 + 10x_7 + 20x_8 &\leq 50 \\
15.8x_1 + 7x_2 + 3.4x_3 + 0x_4 + 0x_5 + 19.1x_6 + 4x_7 + 8x_8 &\geq 15 \\
15.8x_1 + 7x_2 + 3.4x_3 + 0x_4 + 0x_5 + 19.1x_6 + 4x_7 + 8x_8 &\leq 50
\end{aligned}$$

All decision variables are nonnegative. Using Problem Solver, we find that the minimum value of the Z is 1.865, and the optimal values of the decision variables are $x_1 = 0.789$, $x_2 = 0$, $x_3 = 0$, $x_4 = 0.5$, $x_5 = 0$, $x_6 = 0$, $x_7 = 0.193$, and $x_8 = 0.221$ The minimum average cost per day is approximately \$1.87. The values of the decision variables should be interpreted as averages. Slick Rock should order a hamburger about 79% of the time, milk 50% of the time, the garden salad 19.3% of the time, and the chef's salad 22.1% of the time.

If this is not enough food for Slick Rock to feel full, other constraints could be added. For example, he could add the constraint *Each meal will consist of at least 4 servings of something.*

The examples in this section are tiny compared to the linear programming problems routinely solved in industry. Problems with thousands of variables and hundreds of constraints can be satisfactorily solved. The most commonly used method for solving linear programming problems is the simplex algorithm. This is the method used by Problem Solver. See Appendix D for more information.

Exercise Set 3.3

1. Use Problem Solver to solve the following linear programming problems:

   (a) Find the maximum value of $Z = 5x + 6y$ subject to the constraints

$$\begin{aligned} 3x + 4y &\le 10 \\ x + 6y &\le 12 \end{aligned}$$

$$x \ge 0, y \ge 0.$$

   (b) Find the minimum value of $Z = 5x + 6y$ subject to the constraints

$$\begin{aligned} 3x + 4y &\ge 10 \\ x + 6y &\ge 12 \end{aligned}$$

$$x \ge 0, y \ge 0.$$

   (c) Find the minimum value of $Z = 4x - 5y$ subject to the constraints

$$\begin{aligned} 3x + 4y &\ge 10 \\ 6x + 8y &\le 12 \end{aligned}$$

$$x \ge 0, y \ge 0.$$

   (d) Find the maximum value of $Z = -3x + y$ subject to the constraints

$$\begin{aligned} 5x + 7y &\ge 10 \\ -x + 4y &\ge 12 \end{aligned}$$

$$x \ge 0, y \ge 0.$$

2. Use Problem Solver to solve the following linear programming problems:

   (a) Find the maximum value of $Z = x + y + 4z$ subject to the constraints

$$\begin{array}{rcrcrl} x & + & y & + & z & \leq 15 \\ 2x & + & 6y & + & 2z & \leq 40 \end{array}$$

$$x \geq 0, y \geq 0, z \geq 0.$$

   (b) Find the minimum value of $Z = x + 2y + z$ subject to the constraints

$$\begin{array}{rcrcrl} x & + & y & + & 2z & \geq 12 \\ 2x & + & y & + & z & \geq 10 \end{array}$$

$$x \geq 0, y \geq 0, z \geq 0.$$

   (c) Find the maximum value of $Z = x - y + 3z$ subject to the constraints

$$\begin{array}{rcrcrl} x & + & 2y & + & 3z & \geq 14 \\ 2x & + & y & + & z & \leq 10 \end{array}$$

$$x \geq 0, y \geq, z \geq 0.$$

   (d) Find the maximum value of $Z = 3x + 8y + z$ subject to the constraints

$$\begin{array}{rcrcrl} x & + & 7y & + & z & \leq 10 \\ -x & + & 4y & + & 4z & \leq 11 \\ x & + & y & + & 3z & \leq 12 \\ 2x & + & 2y & + & 4z & \leq 14 \end{array}$$

$$x \geq 0, y \geq 0, z \geq 0.$$

3. As Wilbur was applying the third coat of wax to his Phaser GT, he started thinking about what a beauty this car was and how he would invest the \$100,000 that he was sure to make when he sold it. He wanted to invest at least one third of his money in an individual retirement account paying 8% per annum interest. In addition, he wanted to invest some in a growth fund which had about a return of about 11% per year, and some in Ace Manufacturing stocks which had a return of about 18% per year. Since he viewed the Ace Manufacturing stocks as a greater risk, he wanted to invest no more than one fourth of his money on these stocks. On the other hand, since the president of Ace was a good friend, he felt that he should spend at least 5% of his money on Ace stocks. How should Wilbur plan to invest his \$100,000 if he wants to maximize his return while satisfying all of the above constraints?

4. When Mr. Greene is not teaching physics, he can usually be found working in his yard. Rumor has it that several times he has come very close to getting

the "Yard of the Month" honors for his block. He has decided that he will purchase three types of fertilizer for his lawn: ammonium sulfate, urea, and a brand-name mix. The cost per pound for these fertilizers is 10 cents, 12 cents, and 25 cents respectively. Mr. Greene did some careful calculations and concluded that he should apply at least 20 pounds of some nitrogen compound and at least 10 pounds of sulfur. The table below shows how much of these chemicals is contained in each type of fertilizer:

Percent of Nitrogen and Sulfur in Three Fertilizers

Chemical	Ammonium sulfate	Urea	Mix
Nitrogen (%)	21	29	21
Sulfur (%)	0	0	10

In addition, Mr. Greene wants to put at least 10 pounds of ammonium sulfate on his lawn. How much of each type of fertilizer should be used if all of the above requirements are to be met and if the cost is to be minimized?

5. Due to a recent marketing campaign, the Oregon warehouse for Ace Manufacturing has increased its demand to 800 units per day. Solve the problem introduced in Example 3.3.4 with this modification.

6. The Blue Ridge Furniture Company has just discontinued several unprofitable products, and they are considering triple dressers, desks, and coffee tables as replacements. The mill is currently running at 80% capacity, the assembly room at 75% capacity, and the finish room at 70% capacity. The requirements for the three new products for mill work, assembly and finishing are given in the table below:

   Percent of Capacity Used per Item Produced per Hour

	Dresser	Desk	Coffee table
Mill	4	3	2
Assembly	5	4	2
Finish	5	5	3

   If the profit is \$20 on the dresser, \$15 on the desk, and \$12 on the coffee table, how many of each item should be produced per hour in order to maximize the profit?

7. Suppose that Slick Rock put scrambled eggs back into his diet. How much should he eat of each item in order to meet his dietary requirements and at the same time minimize his cost?

8. If Slick Rock added one more item from Burgers 'n Things to his list of possibilities for his diet would you expect his minimum cost to increase or decrease? Give reasons for your answer.

★ 9. Pacific Boats specializes in a 19 foot fiberglass boat hull that they sell to other companies who add motors and accessories and sell them with their own brand label. Pacific Boats has plants in San Francisco, Portland, and Salt Lake City. The boats are shipped from these plants to three warehouses located in Seattle, Los Angeles, and Tampa. The annual requirements of the warehouses, the maximum annual output of the factories and the shipping costs are given below:

Warehouse	Annual Requirement
Los Angeles	4300
Seattle	3500
Tampa	2200

Factory	Maximum Annual Output
Salt Lake City	2200
San Fransisco	4900
Portland	3000

Unit Shipping Costs			
	Los Angeles	Seattle	Tampa
Salt Lake City	70	90	100
San Fransisco	50	80	120
Portland	80	35	135

Find the number of boat hulls that should be shipped from each of the factories to each of the three warehouses.

★★ 10. Northwest Technology builds a two-cycle engine used in a certain snowmobile. They are planning a production schedule for the upcoming year. These motors can be produced in the summer, fall, winter, or spring for sale during any of these seasons. From past experience it is known that the demand is 4500 during the summer, 3500 during the fall, 3000 during the winter, and 1500 during the spring. Up to 3100 can be produced during any one of the seasons. Currently Northwest has 1200 engines in stock and they wish to end the year with at least 500 in stock. The cost of storing an engine is $50 per season, and the cost of building an engine is $800, $750, $800, and $720 for spring, summer, fall, and winter, respectively . Plan a production schedule for next year beginning in the spring that will meet the demand and minimize the cost? *Hint:* This problem can be viewed as a transportation problem in which engines are being transported from the season that they are built into the season that they are sold.

Chapter 4

Probability and Statistics

4.1 Permutations and Combinations

4.1.1 Outfitting Suzie

Suzie, the fashionable poodle, has 3 collars and 2 leashes. We shall calculate the number of different outfits that Suzie has. First let's agree that an outfit consists of both a collar and a leash. Suzie can select a collar in 3 ways. Then for each of the three collar selections she can make one of two choices for a leash. Thus she must have $3 \cdot 2 = 6$ different outfits. This illustrates the following theorem, which is sometimes called the *Fundamental Counting Principle* or the *Multiplication Principle.*

Theorem 1 *If a first task can be carried out in n ways, and if for each of these n ways a second task can be carried out in m ways, then the two tasks can be carried out together in nm ways.*

This principle can be applied to more than two tasks. In fact, it extends to any finite number of tasks. Thus if in addition to her 3 collars and 2 leashes, Suzie also had 4 bows for her hair, she would have $3 \cdot 2 \cdot 4 = 24$ outfits.

Definition 1 *If n objects are placed on a line, then a rearrangement of these objects along the line is called a* permutation *of the objects.*

Products of the form $5 \cdot 4 \cdot 3 \cdot 2 \cdot 1$ come up frequently enough that we use a special notation for them. For any postive integer n, $n!$ (read n factorial) is the product $n(n-1)(n-2)\cdots 1$. We also define $0!$ to be 1. Thus $4! = 4 \cdot 3 \cdot 2 \cdot 1 = 24$.

Theorem 2 *The number of permutations or rearrangements of n objects is given by*

$$n! = n(n-1)(n-2)\cdots 2 \cdot 1.$$

Theorem 3 *The number of permutations of n objects taken r at time is given by*

$$P(n,r) = n(n-1)(n-2)\cdots(n-r+1) = \frac{n!}{(n-r)!}$$

where n and r represent nonnegative integers with $r \leq n$.

Definition 2 *The number of* combinations *of n objects taken r at a time is the number of subsets of size r that can be formed from a set of size n.*

Since any two sets with the same elements are equal, we see that the order in which the elements are listed is not considered when counting combinations. Combinations can be calculated using the following theorem:

Theorem 4 *The number of combinations of n objects taken r at a time is*

$$C(n,r) = \frac{P(n,r)}{r!} = \frac{n!}{r!(n-r)!}$$

where n and r represent nonnegative integers with $r \leq n$.

4.1.2 Problem Solver Example

This example is a demonstration of how to use Problem Solver to calculate $P(10,3)$. There are actually two ways to do this.

Method 1

1. Start Problem Solver and select [Statistics] from the main menu.
2. From the statistics menu select [Probability/counting].
3. From the next menu select [Permutations/combinations].
4. From the permutations/combinations menu select [Total number of objects], and enter 10.
5. Then select [Number of selected objects] and enter 3.
6. Finally select [Permutations].

Method 2

1. Start Problem Solver and select Options from the main menu.

2. Then from the options menu select Calculator. You should now be viewing a rectangle on the screen provided for entering expressions.

3. Type the expression `P(10,3)` and press the Enter key. The value should appear on the screen.

4.1.3 Burgers 'n Things I

Single, double and triple hamburgers are available at Burgers 'n Things. The condiments include catsup, mustard, onions, lettuce, tomatoes, mayonnaise, pickles, and cheese. Slick Rock enjoys these hamburgers and is considering having one each day for lunch. As he relaxed one day after finishing a double hamburger, he decided to try all of the variations of these hamburgers. He resolved to eat one each day for lunch until he had sampled them all.

Slick wondered how long he could eat at Burgers 'n Things without repeating a hamburger. To solve this problem, he reasoned as follows: There are 3 ways to choose the number of hamburger patties. There are 2 ways to choose each of the condiments – either put it on the sandwich or leave it off. Since there are 8 condiments the total number of hamburgers is $3 \cdot 2^8 = 768$. Slick could have a different hamburger every day for over two years.

4.1.4 Burgers 'n Things II

One day as Slick Rock was working his way through the hamburger variations, he was told that three part-time dishwashing jobs were available at Burgers 'n Things. The manager told him that there were 8 applicants who were equally qualified for these positions. Slick Rock wondered how many choices the manager had in hiring 3 of the 8 applicants.

Since the order in which the 3 workers are hired is of no importance, Slick decided that this must be a combinations problem. He concluded that there are $C(8,3) = 56$ ways in which to choose the 3 new workers.

4.1.5 Card Hands

Slick Rock, Wilbur, Maria, and Mr. Greene met each Tuesday night to play cards. As Wilbur was shuffling, Slick wondered out loud how many different 5-card hands could be formed from a standard deck of 52 cards. Maria said, "Since we don't care about the order in which we receive the cards, this is surely a combinations problem. Since we are choosing 5 of 52 cards," she continued, "there are exactly $C(52,5) = 2,598,960$ possible hands."

Not to be outdone, Slick said, "Ah, yes, but how many different five-card hands containing exactly 3 hearts can be formed from a deck of 52 cards?"

Maria, who loved counting problems, quickly responded, "There are $C(13,3) = 286$ ways to choose 3 of the 13 hearts. For each of these 286 ways to choose the hearts, there are $C(39,2) = 741$ ways to select the other 2 cards from the 39 cards that are not hearts. Consequently, there are $C(13,3)C(39,2) = 211,926$ different hands with 3 hearts."

Exercise Set 4.1

1. Use Problem Solver to calculate

   (a) $8!$

   (b) $P(9,4)$

   (c) $C(20,5)$

   (d) $C(5,2)C(9,4)$

2. How many five-card hands containing 2 aces can be formed from a standard deck of 52 cards?

3. In the fall of 1965, Mr. Greene's physics class had 32 chairs and 32 students. How many seating arrangements were possible?

4. Repeat Exercise 3 with 32 chairs and 25 students.

5. The Acme Doll Company has 10 positions available for assembly line workers. Half of these positions are to be filled by women, and half are to be filled by men. If 12 women and 20 men have applied for the positions, in how many ways can the positions be filled?

★ 6. Slick Rock recently purchased an electronic compass for his Phaser GT automobile. Being somewhat of a genius with electronics, he decided to connect this compass to the car's computer system. He was hoping to get directional readouts on his trip computer. Unfortunately, when he was finished his car would travel only north or west. If he attempted to travel south or east, the engine would stop running.

   Slick was nearly late for a date with Jewel and had no time to solve his electrical problem. The town of Mill Creek is laid out in a perfect block pattern of north-south and east-west roads, and Jewel lives 3 blocks north and 6 blocks west of Slick. How many different routes from his house to Jewel's house were available to him?

4.2 Experiments with Equally Likely Outcomes

The set of all possible outcomes for an experiment is called a *sample space* for that experiment. For example, if one tosses a coin, the sample space would be {Heads, Tails}. For a fair coin, these two outcomes are equally likely. If two coins are tossed, then {HH,HT,TH,TT} is a sample space of equally likely outcomes. If we are interested in the number of heads showing, the sample space could be {0,1,2}. These outcomes are not equally likely. The problems in this section use the following theorem:

Theorem 5 *Let S be a finite sample space with equally likely outcomes. If E is an event in S then the probability of E is given by*

$$P(E) = \frac{n(E)}{n(S)}$$

where $n(E)$ is the number of outcomes in E and $n(S)$ is the number of outcomes in S.

4.2.1 Problem Solver Example

Mr. Greene decided to eat 2 chocolate chip cookies before supper in spite of the warning he had received from Mrs. Greene. He quietly slipped into the kitchen, and without looking into the cookie jar, he took two cookies. He knew that the cookie jar contained 5 chocolate chip and 4 peanut butter cookies. As he started to withdraw his hand from the cookie jar, he hesitated for a moment and wondered about the probability of getting 2 chocolate chip cookies.

He reasoned that since he was taking a pair of cookies the sample space must consist of all possible pairs of cookies from the 9 cookies in the jar. There are $C(9,2)$ such pairs. Thus $n(S) = C(9,2)$.

Since he wanted two chocolate chip cookies, E must consist of all possible pairs that could be formed from the 5 chocolate chip cookies. Thus $n(E) = C(5,2)$. Finally,

$$P(E) = \frac{n(E)}{n(S)} = \frac{C(5,2)}{C(9,2)} = \frac{10}{36}.$$

We will solve this type of problem by using the calculator provided in the options menu of Problem Solver.

1. Start Problem Solver and select Options from the main menu.

2. Then select Calculator.

3. Type the expression `C(5,2)/C(9,2)` and press the Enter key. The answer is approximately 0.278.

4.2.2 Lotto

Jimmy Greene, who is a sixth grade student at Mill Creek Elementary, delivers newspapers to earn spending money. His income averages about \$100 per month. One day as Jimmy was folding newspapers in preparation for delivery, he noticed an advertisement for the Washington State Lotto game. In this game, one chooses 6 numbers from the set $\{1, 2, \ldots, 44\}$. The winning 6 numbers are randomly chosen. This means that all sets of 6 numbers have an equal chance of being the winning set. If one guesses all 6 numbers correctly, he or she wins the grand prize, which was two million dollars that week. It cost \$.50 per game to play.

Jimmy saw this as an easy way to instant riches and more free time in the afternoons. He was so excited that he could hardly speak, but he managed to tell his father that he intended to spend \$100 on the game that week.

What ensued could possibly be described as Jimmy's first lesson in probability. Mr. Greene knew that playing the game amounted to selecting a subset of 6 numbers from a set of 44 numbers. Consequently, the sample space for this game is enormous. It consists of all possible subsets of 6 numbers from a set of 44 numbers. There are $C(44, 6) = 7059052$ such subsets. For \$100, Jimmy could play 200 games. Thus $n(E) = 200$, and

$$P(E) = \frac{n(E)}{n(S)} = \frac{200}{C(44, 6)} = \frac{200}{7059052} \approx (2.83)10^{-5}.$$

Mr. Greene subracted this number from 1 and gave Jimmy the probability that he would lose his money. When Jimmy learned that the probability of losing his \$100 was $1 - (2.83)10^{-5} = .9999717$, he decided to spend it on a new bicycle rather than Lotto.

4.2.3 Tenure

The Mill Creek Community College (MCCC) is being sued for sex discrimination. It started last spring when 20 equally qualified professors became eligible for tenure. The MCCC Board of Regents set a limit on tenured positions that allowed only 10 of the 20 candidates to receive tenure. As is customary in tenure decisions, the other 10 would be fired.

Of the 20 candidates for tenure, 11 were women, but of the 10 awarded tenure only 2 were women. One of the 9 women denied tenure sued the MCCC claiming sex discrimination. The MCCC countered all of the candidates were equally likely to receive tenure.

When the case went to trial, the plaintiff's attorney brought in an expert witness to discuss the probability that at most one woman would receive tenure if in fact men and women were equally likely to receive tenure. The sample space for this problem consists of all sets of 10 of the 20 candidates for tenure. This means that $n(S) = C(20, 10)$. Of these sets of 10 professors, $C(9, 9)C(11, 1)$ have one woman and

$C(9,8)C(11,2)$ have two women. Since there are only 9 men, at least one woman must be given tenure. The probability of selecting not more than two women by chance alone is

$$\frac{C(9,9)C(11,1)+C(9,8)C(11,2)}{C(20,10)} = \frac{506}{184756} \approx 0.0027.$$

The jury agreed that this probability is extremely low, and that discrimination did take place. They sided with the plaintiff.

Exercise Set 4.2

1. Use Problem Solver to calculate $C(5,2)/C(21,5)$.

2. Suppose that the Greene's cookie jar now contains 12 chocolate chip and 10 peanut butter cookies. If Mr. Greene draws 4 at random, find the probability that

   (a) all 4 will be chocolate chip.

   (b) none will be chocolate chip.

   (c) exactly 2 will be chocolate chip.

   (d) at least 2 will be chocolate chip.

★ 3. One of the largest employers in Mill Creek is the Inland Empire Insurance Company (IEIC), which is headquartered there. During the past 5 years IEIC has had to lay off 25 office workers and managers. Of the 200 office workers and managers, 40 were over 60 years of age. However, of the 25 workers fired, 22 were over 60 years of age.

   The IEIC now faces a lawsuit for age discrimination. The IEIC adamantly claims that the selections for layoffs were made independent of age and were in fact done randomly. Assume that IEIC is telling the truth, and calculate the probability that under those conditions 22 or more of those fired would be over 60 years of age.

4. Phylicia enrolled in a history class at the Mill Creek Community College. At the end of the term her instructor distributed a review sheet consisting of 50 questions 10 of which would be on the final test.

   In order to make a B in the course Phylicia must get at least 8 of the 10 questions right. One way to be certain of getting at least 8 would be to learn 48 of the 50 questions well enough to get them right on the test.

   After Phylicia had learned 40 of the 50 questions, she reasoned that she had learned 80% of the questions so she should get her B. Find the probability that she will receive a B given that she knows 40 of the 50 questions and assuming that the 10 questions for the test will be chosen randomly.

★★ 5. In the game of poker each player is dealt a hand of 5 cards from a 52-card deck. The deck is divided into 4 suits: clubs, diamonds, hearts, and spades. Each suit has 13 cards: 2, 3, 4, 5, 6, 7, 8, 9, 10, jack, queen, king, and ace. The clubs and spades are black cards, and the hearts and diamonds are red cards. Find the probability of getting each of the following hands:

(a) a hand containing 4 aces

(b) a hand containing all red cards

(c) a hand containing 3 clubs and 2 diamonds

(d) 4 of a kind (4 of the same rank such as 4 kings)

(e) a full house (3 of one kind and 2 of another kind such as 3 kings and 2 aces)

(f) a flush (5 cards of the same suit such as 5 hearts)

(g) a straight (a sequence of 5 cards such as 8-9-10-jack-queen)

(h) a straight flush (a straight from one suit)

(i) a royal flush (a straight flush beginning with a 10 and ending with an ace)

4.3 The Binomial Distribution

An experiment with exactly two outcomes is called a Bernoulli trial. We arbitrarily call one outcome a success and the other a failure. For example, if we toss a coin there are just two possible outcomes: heads and tails. We might designate heads as a failure and tails as a success.

Now consider a sequence of n independent Bernoulli trials where n represents a positive integer. If the probability of success for each trial is p, then the number of successes in all n trials has a binomial distribution. The probability that the number of successes X equals some number k is given by the formula

$$P(X = k) = C(n, k)p^k(1 - p)^{n-k}$$

for $k = 0, 1, \ldots, n$.

4.3.1 Problem Solver Example

Suppose that a fair die is rolled 10 times and that we want to find the probability of getting exactly 2 sixes. If we think of a six as a success and anything else as a failure, rolling the die once is a Bernoulli trial. Rolling it 10 times is a sequence of 10 independent Bernoulli trials since the outcome in one roll has no effect on the outcome in subsequent rolls.

Thus the number of sixes in the 10 rolls has a binomial distribution. We could use the above formula with $p = 1/6$, $n = 10$, and $k = 2$ to calculate the desired probability. The formula could be entered into the calculator of Problem Solver, but there is an easier way.

1. Start Problem Solver, and select [Statistics] from the main menu.
2. Now select [Probability/counting] from the statistics menu.
3. Select [Binomial distribution] from the next menu.
4. Select [Number of trials], and enter 10.
5. Select [Lower limit a:], and enter 2.
6. Select [Upper limit b:], and enter 2.
7. Select [Probability of success in each trial], and enter 1/6.
8. Finally select [Binomial binomial: $P(a \leq X \leq b)$] probability.

Use the following steps if you wish to print your result and if you have not already selected the printer for output:

1. Select [Options] from the main menu.
2. Select [Output device].
3. Select [Printer].
4. Repeat the steps for calculating the binomial probability. This time the results should be printed.

4.3.2 Batter Up

When Slick Rock played baseball with his high school team, the Fighting Weasels, his batting average was 0.3. This means that he got a hit 30% of the times that he was at bat.

Only the final game of his senior year remained. He expected to bat 4 times, and if he could somehow get 3 or more hits he would be inducted into the Weasel Hall of Fame.

As he sat on the bench before that game, he started worrying about his chances of getting the needed hits. To analyze the problem, he assumed that the probability of his getting a hit was 0.3 each time he would go to bat. This produces a sequence of 4 independent Bernoulli trials where the probability of success in each trial is 0.3.

The number of successes X in the 4 trials has a binomial distribution with $n = 4$ and $p = 0.3$. Using Problem Solver, we find that the probability of getting either 3 or 4 hits is

$$P(3 \leq X \leq 4) \approx 0.08.$$

4.3.3 Multiple Choice

One day Mr. Greene gave a 10-question multiple choice test to his physics class. One of the four choices for each question could always be eliminated, but it was tough to choose the correct answer from the remaining three choices. Eddie had forgotten about the test and had not studied. Still he expected to be able to eliminate one choice for each question. Assuming that he guesses among the other three, we shall find the probability that he will make the minimum passing score of 50%.

Consider the sequence of 10 questions to be a sequence of independent Bernoulli trials where the probability of success in each trial is 1/3. Thus the number of questions correct has a binomial distribution with $n = 10$ and $p = 1/3$. Using Problem Solver, we find that the probability of getting at least 5 correct is

$$P(5 \leq X \leq 10) \approx 0.21.$$

Exercise Set 4.3

1. Suppose that a fair coin is flipped 20 times. Find the probability of getting

   (a) exactly 10 heads.

   (b) at least 8 and at most 12 heads.

   (c) all tails.

   (d) at least one head.

2. Mr. Greene once gave a 20-question true-false test to his physics class. His minimum passing score was 10 out of 20. Slick Rock had not studied and found it necessary to guess the answer to each question. In fact, he didn't even read the questions. He said it just confused him to read them. Find the probability that a person using Slick Rock's method will pass the test.

3. A certain drug is known to be effective in 70% of the cases of a rare tropical disease. Dr. Andrews specializes in this disease and regularly prescribes this drug. Find the probability that exactly 5 of his next 7 patients having this desease will be helped by the drug.

4. Three computers are used in a certain spacecraft. The three computers are identical and perform precisely the same functions. The results of each calculation are compared among the three computers as an error check in much the same way that students sometimes compare their answers on assignments. Each computer is capable of controlling the ship, and if one should begin to disagree with the other two, it is turned off.

   The probability that any one computer will fail is 0.001.

   (a) Find the probability that all three computers will fail.

   (b) Find the probability that two of the three computers will fail.

5. A racquetball dropped from 100 inches must bounce at least 68 inches and at most 72 inches. It is known that 95% of the balls produced by a certain machine meet this requirement and 5% do not. Of the next 100 balls find the probability that at least 90 of them will meet the bounce requirement.

6. The probability that a rifle bullet produced by a certain machine will be defective is 0.02. The next 75 that are produced will be tested.

   (a) Find the probability that at least one of them will be defective.

   (b) Find the probability that at least two of them will be defective.

4.4 The Normal Distribution

In the previous sections we have investigated probabilities of events that are made up of a finite number of outcomes. For example, if we toss a coin 3 times there are exactly $2^3 = 8$ possible outcomes. We now consider experiments involving numerical outcomes that can have any value in some interval. For example, if one is asked to name a positive real number that is less than 10, any number in the interval $(0, 10)$ could be given.

Experiments such as measuring the lengths of newborn babies, weighing ripe pumpkins, or recording scores on standardized tests produce results that can be analyzed using the normal distribution. This distribution is completely determined by two numbers: the mean μ and the standard deviation σ. The mean gives the value about which the data tends to cluster. It is the center of the distribution. The standard deviation measures the dispersion of the data from the mean.

Data from a distribution with a large standard deviation ordinarily show more variability than data from a distribution with a small standard deviation. In other words, data from a distribution with a smaller standard deviation tend to be more tightly clustered about the mean than data from a distribution with a large standard deviation.

4.4.1 Problem Solver Example

Slick Rock has just purchased a set of TurboTread radial tires. These tires have a tread life that has a normal distribution with mean $\mu = 40,000$ miles and standard deviation $\sigma = 10,000$ miles. We shall calculate the probability that Slick Rock's tires will last at least 30,000 miles.

1. Start Problem Solver and select [Statistics] from the main menu.
2. Select [Probability/counting] from the statistics menu.
3. Next select [Normal distribution].
4. From the normal distribution menu select [Mean: μ] and enter 40000.
5. Select [Standard deviation: σ], and enter 10000.
6. Select [Lower limit] and enter 30000.
7. Select [Upper limit]. There is no upper limit given so we will use a very large number for that. Five standard deviations above the mean should do nicely. Enter `40000+5*10000`. Problem Solver will do the calculations for you and use 90,000 for the upper limit.
8. Select [Normal probability] for the final answer of approximately 0.84.
9. For a graphical illustration of the probability, select [Draw graph]. This graph is shown in Figure 4.1. The probability that X is between 30000 and 90000 can be interpreted as the area under the normal curve between $x = 30000$ and $x = 90000$.

If you wish to print your results, use the following steps:

1. From the main menu select [Options].
2. Select [Output device].
3. Select [Printer].
4. Return to the normal probability menu and again select [Normal probability].

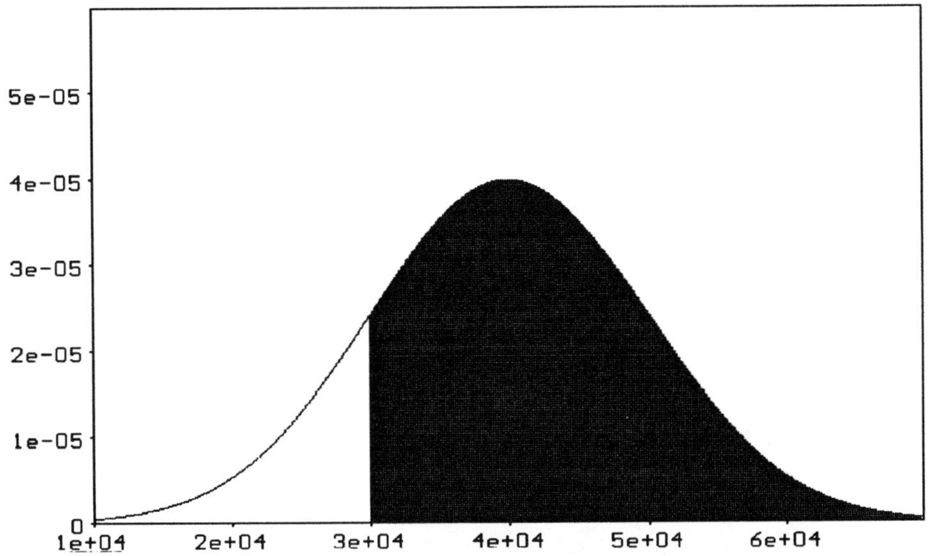

Figure 4.1: Normal Probability and Area

Exercise Set 4.4

1. Let X have a normal distribution with mean $\mu = 50$ and standard deviation $\sigma = 10$. Use Problem Solver to find each of the following:

   (a) $P(30 \leq X \leq 40)$
   (b) $P(40 \leq X \leq 60)$
   (c) $P(X \leq 50)$
   (d) $P(X \leq 45)$
   (e) $P(X \geq 75)$

2. The scores on a standard IQ test have a mean of 100 and a standard deviation of 15. What percentage of those taking the test would be expected to score

   (a) above 100?
   (b) above 140?
   (c) below 95?
   (d) between 80 and 120?
   (e) below 70?

3. The TurboTread Tire Company is planning to introduce a new low priced tire with a 30,000 mile warranty. Tests have shown that these tires have a tread life that approximately follows a normal distribution with mean 35,000 miles and standard deviation 7,000 miles. On what percent of the tires sold should TurboTread expect warranty claims?

4. Over many years of teaching, the final averages in Mr. Greene's physics classes has a normal distribution with mean 75 and standard deviation 12.

   (a) If Mr. Greene requires 90 or better for an A, what percent of his students can expect an A grade?

   (b) If Mr. Greene gives a C to anyone whose average is at least 67 and at most 79, what percent of his students can expect a C grade?

   (c) If Mr. Greene requires at least 53 to pass, what percent of students fail his course?

5. All students applying to Southeastern Washington State University must take ACT tests. The admissions committee studied the data over several years and determined that the test scores have a nearly normal distribution with mean 21 and standard deviation 3. The admissions committee set 17 as a minimum score for any student being considered for admission. What percent of applicants would be rejected based on their ACT test scores?

6. The wieghts of the house cats in Mill Creek have a nearly normal distribution with mean 10.1 pounds and standard deviation 1.9 pounds.

   (a) Approximately what percentage of the cats weigh over 10 pounds?

   (b) Approximately what percentage of the cats weigh less than 5 pounds?

4.5 Organizing Data

Each of the following numbers gives the amount of time (in seconds) required for 100 different cars to accelerate from 0 to 60 mph. Since these data were taken from periodicals directed toward automobile enthusiasts, many of the vehicles tested were high-performance automobiles.

6.0	7.7	8.9	11.1	8.1	5.4	9.5	6.8	8.9	8.0
8.6	7.4	9.5	10.8	7.1	10.4	6.6	6.4	8.0	6.5
7.5	8.9	6.2	7.4	9.9	7.1	8.3	6.7	5.8	6.3
5.1	5.9	6.2	8.6	8.7	9.3	7.5	8.0	10.2	8.6
8.4	10.9	13.1	5.8	10.5	11.3	11.8	9.7	9.5	9.6
6.3	11.9	6.2	7.0	6.9	7.1	13.4	10	14.6	8.5
8.6	10.5	16.1	8.1	14.0	8.0	11.9	7.0	9.2	8.2
6.6	9.8	12.3	10.3	8.8	15.3	8.6	5.5	10.3	9.5
8.8	7.8	4.6	7.8	5.1	7.3	8.3	7.0	9.1	13.7
10.0	9.0	18.7	8.7	9.5	6.2	11.6	8.2	9.6	16.9

Because there are so many numbers, it is difficult to look over the data and get a sense of what they mean. Statisticians have several ways of organizing numbers so that they can be presented in a more understandable way.

Class Interval	Frequency	Relative Frequency
3.95 - 5.95	8	0.08
5.95 - 7.95	28	0.28
7.95 - 9.95	38	0.38
9.95 - 11.95	16	0.16
11.95 - 13.95	4	0.04
13.95 - 15.95	3	0.03
15.95 - 17.95	2	0.02
17.95 - 19.95	1	0.01

Table 4.1: Elapsed Time for 0-60 mph for 100 Automobiles

4.5.1 Tables and Graphs

Consider separating the above numbers into classes or intervals. These intervals will be called *class intervals*. Since the smallest number is 4.6 and the largest is 18.7, we will use the intervals $[3.95, 5.95], [5.95, 7.95], \ldots, [17.95, 19.95]$.

The selection of 3.95 as the left end-point and 2 as the width was somewhat arbitrary. However, we do want all of the data included, so the left end-point of the first interval certainly should be less than 4.6. Also, since the intervals share end-points, we want to avoid the ambiguity of having any data value fall on an end-point of an interval. This is why the end-points are given with one more decimal place of precision than the data.

With a width of 2, we need 8 class intervals to include all of the data. We could use a shorter width and more intervals or fewer and wider intervals. Too many intervals defeat the purpose of grouping the data, and too few intervals mask too much of the information contained in the data.

With the above intervals we can form a relative frequency table as shown in Table 4.1. The frequency in each interval is the number of data values falling in that interval. The relative frequency is the frequency divided by the total number of data values.

The relative frequency table can easily be converted to a graph called a histogram as shown in Figure 4.2. A histogram is a form of a bar chart. The left and right boundaries of each bar are determined by the end-points of the class intervals. The height of each bar is given by the relative frequency of data falling into the corresponding class interval.

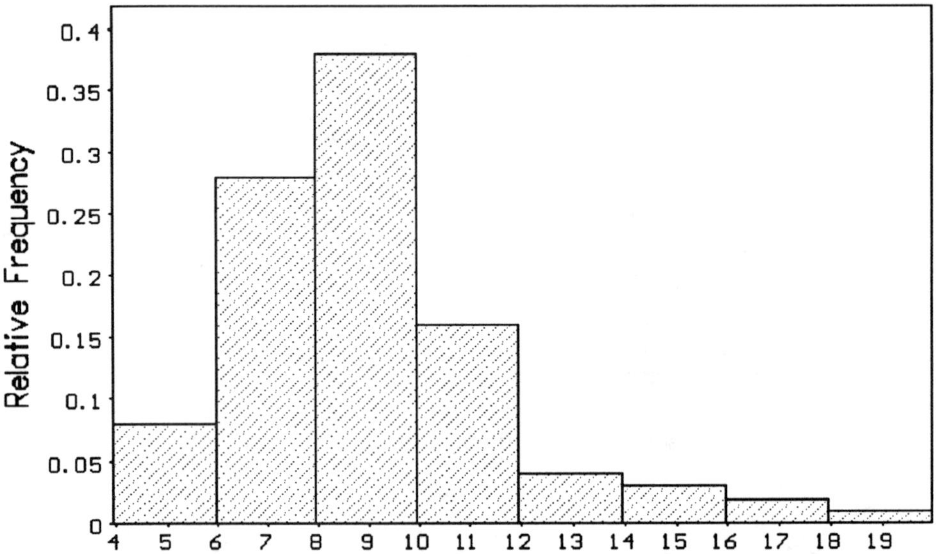

Figure 4.2: Elapsed Time for 0-60 mph for 100 Automobiles

4.5.2 The Mean, Median, and Standard Deviation

It is possible to collapse all of the 0-60 mph data from the last section into a single number. The average or *mean* of the numbers $x_1, x_2, \ldots, x_n$ is given by the formula

$$\overline{x} = \frac{x_1 + x_2 + \cdots + x_n}{n}.$$

The mean of the 0-60 mph data is 8.929 seconds.

The mean is an estimate of the center of the distribution of numbers. The *median* is another measure of this center. To calculate a median, first arrange the data values in increasing order. If the number of data values is odd, the median is the middle value. Otherwise there is no single middle value, and the median is the mean of the two middle values. For small data sets, the median is easy to calculate. For large data sets, arranging the numbers in increasing order is harder than adding the numbers, so the mean is easier to calculate. The median of the 0-60 mph data is 8.6 seconds.

The *standard deviation* is a measure of dispersion from the mean. Widely scattered data will have a larger standard deviation than data that are more tightly clustered about the mean. The standard deviation of the numbers $x_1, x_2, \ldots, x_n$ is given by the formula

$$s = \sqrt{\frac{(x_1 - \overline{x})^2 + (x_2 - \overline{x})^2 + \cdots + (x_n - \overline{x})^n}{n-1}}.$$

The standard deviation of the 0-60 mph data is 2.63 seconds.

4.5.3 Problem Solver Example

We will show the details of getting Problem Solver to find the mean, median, and standard deviation for the 0-60 mph data. In addition, we will generate a relative frequency table and a histogram.

1. Start Problem Solver and select [Statistics] from the main menu.
2. From the next menu select [Descriptive statistics].
3. Select [Enter/edit data]. There are 100 data values. Enter `100` when this number is requested.
4. You should be viewing the edit screen. Enter the data values. When you press the Enter key, the cursor will automatically move to the correct position for the next data entry. If you need to manually control the cursor, use the arrow keys.
5. After the data have been entered, press the Esc key to leave the editor. Now select [Mean/median/standard deviation]. These values will be displayed.
6. Select [Frequency table].
7. Select [Change table specifications].
8. Select [Number of class intervals] and enter 8.
9. Select [Width of class intervals] and enter 2.
10. Select [Left end-point of first interval] and enter 3.95.
11. Leave this menu by pressing the Esc key.
12. Select [Display table].
13. Press the Esc key twice to return to the descriptive statistics menu.
14. To see a histogram select [Histogram] and then [Draw histogram].

To print your results select [Options] then [Output device] and finally [Printer]. Then return to the statistics menu and again request the mean, the frequency table, or whatever you want to have printed.

Exercise Set 4.5

1. Use Problem Solver to find the mean, median, and standard deviation for each of the following data sets. Comment on your results.

   (a) 8,8,8,8,8,8,8

   (b) 7,8,8,8,8,8,9

   (c) 0,0,0,8,8,8,32

2. The annual salaries at Edgar's Sheet Metal Shop are \$10,000, \$12,000, \$12,000, \$13,000, \$12,500, \$13,500, and \$105,000. Edgar's five employees earn the first five salaries listed, and Edgar pays himself the last salary.

   (a) Find the mean and median of these salaries.

   (b) Edgar uses the mean as a recruiting tool for new employees. Is this fair?

   (c) If a labor union representative tried to organize the employees, would he or she be more likely to use the mean or the median in his or her campaign to get the workers to join the union?

   (d) Make a general suggestion about choosing between the median and mean as an estimate of the center of a distribution of number.

3. The following numbers represent the expected city-driving gasoline mileage (in miles per gallon) for a group of cars tested by a popular automobile magazine. Construct a relative frequency table and a histogram for this data.

16	21	19	18	20	13	17	28	17	22
22	22	21	29	23	17	20	21	20	23
19	20	18	17	25	24	19	19	20	18
23	22	19	17	16	20	25	20	16	19
19	18	15	17	25	26	21	20	18	34
21	20	28	29	22	22	25	25	19	24

4. Table 4.2 gives data for 18 ski areas in Washington:

   (a) Find the mean vertical drop for these ski resorts.

   (b) Construct a histogram for the vertical drop.

   (c) Construct a histogram for the number of lifts.

Name	Vertical Drop (ft.)	Number of Lifts	Name	Vertical Drop (ft.)	Number of Lifts
Aplental	2200	9	Mount Spokane	1800	5
Crystal Mountain	3100	9	Pacific West	1143	5
Echo Valley	500	4	Sitzmark	650	3
49 Degrees North	1850	4	Ski Acres	1020	13
Hurricane Ridge	660	3	Ski Bluewood	1125	3
Leavenworth Ski Bowl	600	2	Snoqualmie	900	11
Loup Loup	1240	4	Squilchuck Ski Bowl	500	2
Mission Ridge	2140	6	Stevens Pass	1800	9
Mount Baker	1500	9	White Pass	1500	6

Table 4.2: Washington Ski Areas

4.6 Linear Regression

If the graph of a set of ordered pairs of the form (x, y) falls nearly on a line, we say that x and y are nearly linearly related. We call y the dependent variable, and we call x the independent variable. To find the line with equation $y = a + bx$ that best fits the ordered pairs $(x_1, y_1), (x_2, y_2), \ldots, (x_n, y_n)$ calculus techniques are used to select the coefficients a and b so that the expression

$$(a + bx_1 - y_1)^2 + (a + bx_2 - y_2)^2 + \cdots + (a + bx_n - y_n)^2$$

is as small as possible. Note that the expression $a + bx_i - y_i$ is the difference between the y-coordinate of (x_i, y_i) and the y-coordinate corresponding to $x = x_i$ on the line $y = a + bx$. We are thus choosing a and b to minimize the sum of the squares of the errors. This line is sometimes called best *least squares* fit to the data. It is also called the *regression line.*

4.6.1 Problem Solver Example

The data given in Table 4.3 give the weight and expected city gas mileage for 25 automobiles. We shall use Problem Solver to find and graph the regression line for these data treating the weights as the independent variable and the fuel economy as the dependent variable. In addition, we shall use the regression line to predict the mileage of a car weighing exactly 2500 lb.

1. Start Problem Solver and select [Statistics].

2. Select [Linear regression].

3. Select [Enter/edit data]. There are 25 pairs of data values to enter. Enter 25 when this number is requested.

Weight	mpg	Weight	mpg
2380.0	29.0	2256.0	19.0
3510.0	13.0	2450.0	20.0
3450.0	18.0	1840.0	32.0
2560.0	21.0	2310.0	23.0
2728.0	22.0	2725.0	18.0
2498.0	19.0	2700.0	23.0
3295.0	16.0	3110.0	23.0
2920.0	20.0	3100.0	17.0
3375.0	17.0	2908.0	20.0
3167.0	14.0	2811.0	19.0
2860.0	24.0	3271.0	17.0
2060.0	25.0	3754.0	19.0
2200.0	35.0		

Table 4.2: Gas Mileage and Curb Weight

4. You should be viewing the matrix editor now. Enter the weights in the first column and the gas mileage in the second column. Enter all 25 ordered pairs of data, and then press the Esc key to leave the editor.

5. Select [Compute regression line]. The equation $y = -0.00750687x + 42.0107$ will be displayed.

6. Select [Graph regression line].

7. From the next menu again select [Graph regression line]. The graph should show both the data and the regression line.

8. Press the Esc key four times to return to the main menu.

9. Select [Functions] and then [Functional values].

10. You need to select your function from a list of functions. One of them (probably the second function) should be formed from the equation of the regression line that was just calculated. Select that function.

11. Select [Number of functional values] and enter 1.

12. Select [Enter x-values from keyboard] and enter 2500. Press the Esc key to leave the matrix editor.

13. Select [Calculate functional values].

Serving Weight (g)	Calories
147	353
197	455
184	493
234	610
197	375
156	292
234	501

Table 4.4: Serving Weight and Calories

14. The expected fuel economy of approximately 23.2 mpg. corresponding to 2500 lb should appear in the second column.

15. To print any of your results select [Options]. Then select [Output device] and [Printer]. Once again request the equation of the regression line, a graph of the regression line, or whatever you want to have printed.

Exercise Set 4.6

1. Find the equation and the graph of the regression line for each of the following data sets:

   (a) $(1,2),(3,3)$

   (b) $(1,2),(3,3),(5,5)$

   (c) $(1,2),(3,3),(5,5),(7,6)$

2. Find the equation of the line that passes through the two points (3,-4) and (5,1).

3. Use the data in Table 4.2 to study the relationship between the vertical drop and the number of chair lifts at the listed ski resorts. Treat the vertical drop as the independent variable and the number chair lifts as the dependent variable. Find and graph the regression line along with the ordered pairs of data.

4. A fast food chain published a guide to its menus that includes the weight of each item served along with the number of calories and other nutritional information. The serving weights and calorie content for six different sandwiches are listed in Table 4.4.

(a) Using serving weight as the independent variable and calorie content as the dependent variable, find the equation of the regression line for this data.

(b) Create a graph of data and the regression line.

(c) Predict the number of calories for a 200g item.

4.7 Simulation

Problem Solver can generate random data sets with various distributions. This makes it possible to do some interesting simulation problems.

4.7.1 Problem Solver Example

When Wilbur was in the eighth grade, he never found time to study. His teacher liked to give true-false tests, and Wilbur usually guessed the answers without even reading the questions. During the year, 30 true-false quizzes were given. Each quiz had 10 questions and was scored on a basis of 10 points. We shall use Problem Solver to simulate the 30 quizzes and calculate Wilbur's average score.

1. Start Problem Solver, and select [Statistics].

2. Select [Descriptive statistics].

3. Select [Generate random data].

4. Since Wilbur didn't even read the questions, each quiz was a sequence of 10 independent Bernoulli trials where the probability of success in any trial was 0.5. Consequently the total number of questions answered correctly on any quiz has a binomial distribution. Select [Binomial distribution].

5. The total number of trials is 10, the number of questions on each quiz. Select [Total number of trials], and enter 10.

6. Select [Probability of success in each trial], and enter 0.5.

7. Select [Number of random data points to generate], and enter 30, one for each quiz.

8. Select [Generate points]. The 30 numbers will be displayed. Each represents a quiz score. Because these scores are randomly generated, you should not expect to get exactly the same results twice.

9. Select [Mean/median/standard deviation]. The mean is Wilbur's average score.

Exercise Set 4.7

1. Suppose that a class of 300 students at a large university took a 10-point true-false quiz. Further suppose that these students had studied special techniques for taking this type of test, and that even though none knew anything about the course, they could answer the questions correctly with probability 0.7.

   (a) Generate a set of 300 numbers that simulate the scores on this quiz.

   (b) Find the mean quiz score.

   (c) Create a histogram to display the distribution of quiz scores.

2. When Maria played softball for the Weaselettes at Mill Creek High School, her batting average was 0.256. During a typical season, she would go to bat 25 times. Create a set of numbers that simulate the number of hits that she could expect to get in a year. Do this for 4 years.

3. Suppose that 500 married couples will have 3 children each. Assume that boys and girls are equally likely.

   (a) Generate a set of numbers that will represent the number of girls in each of the 500 families.

   (b) Display your results in a histogram.

4. The Mill Creek Mudstomper boot lasts an average of 22 months with a standard deviation of 6 months.

   (a) Use the normal distribution to simulate the life of 100 pairs of these boots.

   (b) Display your results in a histogram.

5. Mom's Bakery is famous for its large chocolate chip cookie called the Buffalo Chip. The number of chocolate chips in these tasty cookies has a normal distribution with mean 8.1 and standard deviation 1.9.

   (a) Use the normal distribution to simulate the testing of 200 of these cookies.

   (b) Display your results in a histogram.

   (c) Find the mean and standard deviation of the number of chocolate chips found in your sample of 200 of these cookies.

6. While Wilbur was in college, he earned money in the summer by cutting firewood and selling it by the cord. The "cords" that he sold were actually 0.9 cords on the average with a standard deviation of 0.1 cords.

   (a) Use the normal distribution to simulate the sales of 20 cords of wood.

(b) Display your results in a histogram.

(c) How many of the 20 "cords" were actually less that a full cord?

7. Each time a certain secret agent accepts a mission, the probability that she will succeed is 0.95. She has just accepted a string of 10 missions.

   (a) Use the binomial distribution to model the number of successful missions in the 10 that she accepted.

   (b) Use the binomial distribution to model the number of successful missions in her next 100 missions.

8. A new machine produces bolts with a nominal length of 6 inches. The lengths of these bolts have a normal distribution with mean 6 and standard deviation 0.01 inches.

   (a) Use the normal distribution to model the lengths of the next batch of 100 bolts.

   (b) Display your results in a histogram.

   (c) Give the mean and standard deviation for this batch of 100 bolts.

9. A group of 100 students participated in an experiment where each student flipped a coin 5 times. Use the binomial distribution to model this experiment.

   (a) How many students got heads 3 or more times?

   (b) How many students got all tails?

   (c) How many students got exactly one head?

   (d) Graph the results in a histogram with 4 class intervals (for 0 heads, 1 head, 2 heads, and 3 heads).

Chapter 5

Functions, Models, and Graphs II

5.1 Sequences

Definition 3 *An* infinite sequence *is a function whose domain is the set of positive integers.*

If f is such a function, then the range of f is the set $\{f(1), f(2), f(3)...\}$. When we speak informally of a sequence we are usually thinking of the list of values in the range of the sequence. We denote this list by $a_1, a_2, a_3, \ldots$, where $a_i = f(i)$ for each positive integer i.

5.1.1 Problem Solver Example

One warm day early in the spring, Mr. Greene considered the problem of getting his son Jimmy to mow the lawn. Jimmy loved to mow the lawn the first six times or so each season, but after that it usually became increasingly difficult get him behind the lawn mower. Mr. Greene knew that he needed a clever scheme to encourage Jimmy near the end of the mowing season. He told Jimmy that he would start the season at a very low wage. He would earn only one penny the first time he cut the lawn. This didn't bother Jimmy since he would gladly have cut it without pay the first time. Mr. Greene went on to say that Jimmy's wage would double each time he cut the lawn over the next 14 weeks. Mr. Greene was hoping that in the weeks ahead he could offset Jimmy's decreasing interest with a sharply increasing level of greed.

Jimmy knew that he could cut the lawn once each week for up to 14 weeks. He also knew that if he missed a week he would also miss one doubling of his wage. Jimmy began to wonder just how good a deal this was.

We shall use Problem Solver find his wage for each of the 14 weeks.

1. Load Problem Solver and select Functions.

2. From the functions menu select Sequences.

3. Select Number of terms and set this value equal to 14.

4. Select Calculate terms. In the next screen $x[i]$ denotes Jimmy's wage the ith time that he cuts the lawn. The default choice for $x[1]$, his wage for the first cutting, is 1. Since this is the correct value, we need not change it.

5. We need to enter the formula that calculates Jimmy's wages for weeks 2 through 14. Select the menu line containing the formula for $x[i]$, and enter `2*x[i-1]`. With this formula we are calculating the wage for any particular cutting by doubling the wage for the previous cutting. We could also have entered `2^(i-1)`. Both expressions give the same result.

6. Finally select Calculate x-values. The wages (in cents) will be displayed on the screen. We can see that Jimmy will earn $81.92 for his last cutting. We can also see that if Jimmy skips one week during the summer, it will cost him $81.92.

7. To print the results press the Esc key to return to the sequences menu. Select Options. Then select Output device and finally select Printer. After you have chosen the printer as the output device, press the Esc key to return to the sequences menu. Select Calculate terms and then Calculate x-values. This time the results will be printed.

In this sequence each term can be found by multiplying the preceeding term by a fixed constant, in this case 2. Sequences that can be formed in this manner are called *geometric progressions.*

5.1.2 Wilbur's Savings Plan

Wilbur placed $1,500 into a savings account that paid 6.1% per annum interest compounded monthly. He plans to add no more money to this account, and he wants to know how much money will be in the account, including interest, at the end of each month for the next 24 months.

The compound interest formula is

$$P = A(1 + r)^n$$

where A represents the amount deposited, r represents the rate of interest per compounding period, and P represents the amount in the account after n compounding periods. For Wilbur's bank account, $A = 1500$ and $r = 0.061/12$. Thus the amount in the account after n months is given by the formula

$$P = 1500\left(1 + \frac{0.061}{12}\right)^n.$$

Using Problem Solver, we can calculate 24 terms of the sequence whose ith term gives the amount in the account after i months. This sequence is defined in Problem Solver by the two equations

$$\begin{aligned} x_1 &= 1500(1 + 0.061/12) \\ x_i &= 1500(1 + 0.061/12)^i. \end{aligned}$$

Note that Problem Solver requires that the first term of the sequence be explicitly given. The terms of this sequence are given in the table below:

Month	Amount	Month	Amount	Month	Amount
1	1507.63	9	1570.04	17	1635.03
2	1515.29	10	1578.02	18	1643.34
3	1522.99	11	1586.04	19	1651.70
4	1530.73	12	1594.10	20	1660.09
5	1538.52	13	1602.21	21	1668.53
6	1546.34	14	1610.35	22	1677.02
7	1554.20	15	1618.54	23	1685.54
8	1562.10	16	1626.76	24	1694.11

Exercise Set 5.1

1. Use Problem Solver to find the first 50 terms of the sequence $3, 5, 7, 9, \ldots$. Sequences such as this one in which each term can be found by adding a fixed constant to the preceeding term are called *arithmetic progressions.*

2. Use Problem Solver to find the first 50 terms of the sequence

   $\frac{2}{3} + 1, \left(\frac{2}{3}\right)^2 + 2, \left(\frac{2}{3}\right)^3 + 3, \ldots$

3. Suppose Jimmy was paid 1/3 cent for the first mowing and that his wage tripled for each mowing after the first. How much would he make the 14th time he mowed the lawn?

4. At the end of the summer, Jimmy placed \$500 in a 6-month certificate of deposit that paid 7% per annum compounded daily. Find the amount in the account for each of the 180 days that the bank will be holding Jimmy's money.

★ 5. Maria borrowed $125,000 to buy a house. Her interest rate is 11% per annum, and it is always calculated on the declining balance, the amount Maria still owes. If Maria agreed to make monthly payments of $1,300, find the amount that she would owe at the end of each month for the first two years of this loan. Assume that the first payment is due exactly one month after she gets the loan. *Hint*: If x_i represents the amount that Maria owes at the end of month i after making her payment, you can write an equation relating x_i to x_{i-1}. Remember that the amount she owes at the end of a month includes the balance from last month plus the interest on that balance for a month minus the current payment. Tables giving the declining balance of a loan are called *amortization tables.*

5.2 Series

Definition 4 *An expression of the form* $a_1 + a_2 + a_3 + \cdots$ *is called an* infinite series.

A series is the sum of the terms of a sequence. For example, corresponding to the sequence $1, 1/2, 1/4, \ldots$, we have the related series $1 + 1/2 + 1/4 + \cdots$. This type of sum is interesting because it involves infinitely many terms.

In order to study a series, we shall look at the sum of the first n terms. Such sums are called *partial sums.* Consider the above example. The partial sum of the first 2 terms is 3/2. The partial sum of the first 3 terms is 7/4. The following table gives partial sums of n terms for several values of n.

Number of terms	Partial sum		
4	$1 + 1/2 + 1/4 + 1/8$	=	15/8
5	$1 + 1/2 + 1/4 + 1/8 + 1/16$	=	31/16
6	$1 + 1/2 + \cdots + 1/32$	=	61/32
7	$1 + 1/2 + \cdots + 1/64$	=	127/64

The partial sums appear to be getting close to 2. When the partial sums get closer and closer to some number, we say that the sum of the series is that number. Here it appears that the sum of the series is 2. This series is an example of a geometric series, and it can be shown that our guess is correct; the sum is 2.

5.2.1 Problem Solver Example

We shall use Problem Solver to estimate the sum of the series given above. Notice that the partial sums form a sequence. It is this observation that will allow us to solve the problem. We will use the sequence capabilities of Problem Solver.

1. Start Problem Solver and select Functions.

2. Select Sequences.

3. Now select [Number of terms] and enter 100.

4. Select [Calculate terms].

5. Select [x[1]] and enter 1.

6. At this stage we need a formula for $x[i]$, the partial sum of the first i terms of the sequence. The terms of the sequence $1, 1/2, 1/4, \ldots$ could be written in the following way:

first term:	$(1/2)^0$
second term:	$(1/2)^1$
third term:	$(1/2)^2$
ith term:	$(1/2)^{i-1}$

Now select the line that gives the equation whose left side is `x[i]` and enter `x[i-1]+(1/2)^(i-1)`. In this way, each $x[i]$ is the sum of the first $i-1$ terms of the sequence plus the ith term. This means that x_i is the sum of the first i terms.

7. Select [Calculate terms]. Partial sums of $1, 2, \ldots, 100$ terms should appear on the screen. Are these sums approaching 2?

Exercise Set 5.2

1. Consider the series

$$1 + \left(\frac{1}{2}\right)^2 + \cdots + \left(\frac{1}{n}\right)^2 + \cdots$$

   (a) Find the partial sum of the first

      i. 5 terms

      ii. 20 terms

      iii. 100 terms

      iv. 200 terms

   (b) Are the partial sums approaching a particular number?

2. Use Problem Solver to find the sum

$$\frac{1}{1+1^2} + \frac{1}{1+2^2} + \frac{1}{1+2^3} + \cdots + \frac{1}{1+2^{20}}.$$

3. Consider the example of Wilbur's savings account given in the last section. Wilbur deposited \$1,500 into an account that paid 6.1% per annum compounded monthly. Find the sequence of monthly interest payments that the bank will be adding to Wilbur's account over the next 24 months. Then add these amounts together to get the total interest paid to Wilbur. Is there an easier way to do this problem?

★ 4. Maria purchased a stereo system for \$1,500 on credit. She is to repay this 15% loan over a 3-year period.

   (a) The formula

   $$P = A\frac{r}{(1-(1+r)^{-n}}$$

   can be used to calculate the payment P for a loan amount A where the interest rate is $100r\%$ per payment period, and the loan is to be repaid in n payments. Use the above formula and the calculator in Problem Solver's options menu to calculate Maria's monthly payment.

   (b) After making 13 payments, Maria wanted to pay the entire remaining balance with her 14th payment. The formula

   $$B = P\frac{(1+r)^{n-1}}{r}$$

   gives the outstanding balance when n payments remain to be paid. Use this formula to determine how much Maria will have to pay in her 14th payment.

   (c) Use a partial sum of a series to add the interest Maria pays each month and find the total amount of interest that she will pay during the loan. Is there an easier way to do this problem? If you can think of an easier way, solve the problem again using the easier method.

5.3 Exponential and Logarithmic Functions

5.3.1 Exponential Functions

Exponential functions are defined by equations of the form $f(x) = a^x$ for all real x, where a is a positive constant not equal to 1. For example, $f(x) = 2^x$ and $g(x) = (1/3)^x$ are exponential functions. The natural exponential function uses base $e = 2.71828\ldots$ and is sometimes denoted by $\exp x$. The Problem Solver uses the notation `exp(x)`.

5.3.2 Problem Solver Example

The population $P(t)$ of a certain city t years after 1980 is given by

$$P(t) = 50000e^{0.005t}.$$

We shall predict the population in 1995, find the time when the population will be 60,000, and graph the population from 1980 to 2080.

1. Start Problem Solver and select Functions from the main menu.
2. Select Variable name and enter t for the variable name.
3. Select Edit/enter functions and then select the first function.
4. Type the expression `50000*exp(0.005*t)` and press the Enter key.
5. Press the Esc key to return to the functions menu.
6. Select Functional values.
7. From the list of functions select the function that you just entered. This will bring up the functional values menu.
8. Select Number of values, and enter 1.
9. Select Enter x-values from keyboard. When the matrix editor appears on the screen, enter 10 (since 1990 is 10 years after 1980).
10. Press the Esc key to leave the matrix editor.
11. Select Calculate functional values. The prediction for 1990 should be approximately 52,564.
12. Press the Esc key twice to return to the functions menu.
13. Select Solve $f(t) = 0$.
14. Again select the function that you entered.
15. Now select your function from the top line of the equation solver menu. Since the equation solver handles equations of the form $f(x) = 0$, and our equation is of the form $f(x) = 60,000$, we must subtract 60,000 from our function. When your function appears in the editor, press the End key to move the cursor to the end of the function. Type `-60000`, and press the Enter key. Your new function should be `50000*exp(0.005*t)-60000`.
16. Select Initial guesses, and enter 0 and 100.

17. Select [Compute solution]. You will see that it will take about 36.46 years for the population to reach 60,000. This should happen in the year 2016.

18. Before leaving the equation solver, again select the first line of the menu to edit your function. Delete the `-60000` that you entered for the purpose of solving the equation.

19. Press the Esc key to return to the functions menu. Now select [Graph functions].

20. Select [Function selection] and then select the function that you entered. Press the Esc key to return to the graphing menu.

21. Select [Change graph specifications] and then select the population function.

22. Select [Minimum x-coordinate on the axis] and enter 0.

23. Select [Maximum x-coordinate to graph] and enter 100.

24. Select [Minimum y-coordinate on axis] and enter 40000.

25. Select [Maximum y-coordinate on axis] and enter 100000.

26. Press the Esc key to return to the graphing menu.

27. Finally select [Draw graph].

28. To print your results select [Options] followed by [Output device] and [Printer]. Then repeat any of the above steps leading to results that should be printed.

5.3.3 Radiocarbon Dating and Nuclear Waste

Carbon-14 is a radioactive isotope of carbon with a half-life of about 5,770 years. This means that if one begins with a certain quantity of carbon-14, half of it will decay into other elements in 5,770 years. During the next 5,770 years, another half will decay, leaving only one fourth of the original amount. Most of the carbon occurring naturally is carbon-12, but carbon-14 is present in small quantities in the atmosphere. Living plants and animals contain the same ratio of carbon-14 to carbon-12 as does the atmosphere. At the time of death, the carbon-14 begins to decay, and the ratio of carbon-14 to carbon-12 decreases. This idea is used to date fossils. If only one fourth the expected carbon-14 is present in a piece of petrified wood, it is assumed that the tree died about $2 \cdot 5770 = 11,540$ years ago.

We could use the same analysis to see how long nuclear waste will be dangerous. If A_0 units (pounds could be used) of a raDioactive material with a half-life of L are present now, the amount present after t years is given by

$$A(t) = A_0 \left(\frac{1}{2}\right)^{\frac{t}{L}}.$$

5.3.4 Logarithmic Functions

A logarithmic function is written in the form $f(x) = \log_a x$ for $x > 0$ where the base a is a positive constant not equal to 1. Logarithmic functions and exponential functions are related by the equivalence:

$$y = \log_a x \text{ if and only if } a^y = x.$$

Thus $\log_2 8 = 3$ since $2^3 = 8$.

The natural logarithm uses base e and is denoted by $\ln x$. The common logarithm uses base 10 and is denoted by $\log x$.

5.3.5 Decibels, the Richter Scale, and Jimmy Greene

Much to the annoyance of Mr. Greene, his son Jimmy had the family stereo set at a very high volume. After a while, Jimmy's favorite song came on, and he turned up the stereo even more to what could be described only as a truly deafening level. When this happened, Mr. Greene walked into the living room and said loudly enough to be heard, "Jimmy, the stereo was already too loud and you just increased the sound intensity 100-fold. Now turn down the volume."

"You don't have to yell," shouted Jimmy. "I did not increase the sound by a factor of 100. My meter showed that I increased the sound only from 95 to 115 decibels. How is that increasing the intensity by 100-fold?"

Mr. Greene explained to Jimmy that the decibel level D of a sound of intensity I is given by the logarithmic equation

$$D = 10 \log \frac{I}{I_0}$$

where I_0 represents a sound intensity just at the threshold of hearing. First Mr. Greene divided both sides of the equation by 10 to get

$$\frac{D}{10} = \log \frac{I}{I_0}.$$

Writing this equation in exponential form, Mr. Greene showed that

$$10^{D/10} = \frac{I}{I_0}.$$

Multiplying both sides by 100 or 10^2 produced the equation

$$10^{(D/10)+2} = 100\frac{I}{I_0}.$$

Converting this back into the logarithmic form led to

$$\frac{D}{10} + 2 = \log \frac{100I}{I_0}.$$

Finally multiplying by 10 yielded the equation that Mr. Greene wanted:

$$D + 20 = 10 \log \frac{100I}{I_0}.$$

Upon seeing this equation, Jimmy was finally convinced that increasing the decibel level D by 20 does increase the sound intensity I by a factor of 10^2. In fact, each 10 units of increase on the decibel scale corresponds to a 10-fold increase in the sound intensity.

"The Richter scale used to measure earthquakes works much in the same way," continued Mr. Greene. "The Richter-scale rating of a quake of intensity I is given by the logarithmic equation

$$R = \log \frac{I}{I_0}$$

where I_0 is the intensity of a very small quake. In this case, the factor of 10 is missing. Consequently, an increase of 1 in the Richter scale rating corresponds to a 10-fold increase in the intensity of the quake. Thus a quake measuring 7.8 on the Richter scale has an intensity 10 times greater than one measuring 6.8 on the Richter scale."

Exercise Set 5.3

1. Use Problem Solver to graph each of the following functions: In each case, select appropriate intervals on the axes, so that your graph shows enough to give you an idea of what the whole graph looks like.

   (a) $f(x) = x \log(x - 1)$

   (b) $g(x) = \exp(x) - \exp(-x)$

   (c) $h(x) = \exp(\ln x)$

   (d) $u(x) = \ln |x|$

   (e) $c(x) = 2 - 0.5e^{-x}$

2. Suppose that you are asked to store 112 pounds of a deadly radioactive material with a half-life of 300 years. You have been given special storage barrels expected not to leak for 1,000 years, and you are to bury the radioactive material in these barrels.

   (a) Find the amount of radioactive material that could enter the environment if the barrels leak after 1,000 years.

   (b) Find the amount of radioactive material in the barrels after $100, 200, \ldots, 1000$ years.

(c) After how many years will there be 85 pounds of radioactive material in the barrels?

(d) Graph the amount of radioactive material in the barrels over the next 1,000 years.

3. When Jimmy was born, Mr. Greene purchased a \$2,000 savings certificate hoping that it would cover the cost of Jimmy's college education. The interest rate on this certificate was 8% per year compounded annually. The amount $A(t)$ in the account after t years is given by the formula

$$A(t) = 2000(1 + 0.08)^t.$$

(a) Find the amount in the account at the end of each year for 17 years.

(b) How long did it take for the account to reach \$3,500?

(c) Graph the amount in the account for 17 years.

4. The city of Compton had a population of 25,000 in 1985. Its population t years after 1985 is given by the equation

$$P(t) = 25000e^{0.05t}.$$

(a) What will the population be in 2015?

(b) When will the population reach 40,000?

(c) Graph the population for the next 50 years.

5. Burgers 'n Things has discovered that the function

$$B(t) = 100 - 80e^{-.5t}$$

gives the number of hamburgers per hour that their employees produce after having been on the job for t days. This is an example of a *learning curve.*

(a) Graph this function for $0 \leq t \leq 10$.

(b) How long will it take for an employee to be able to produce 50 hamburgers per hour?

(c) How many hamburgers per hour can an employee produce after 6 days on the job?

6. Find $\frac{I}{I_0}$ for a sound of

(a) 30 decibels.

(b) 70 decibels.

(c) 100 decibels.

(d) 150 decibels.

(e) Graph D as a function of I/I_0.

7. Find $\frac{I}{I_0}$ for an earthquake with the following ratings on the Richter scale: 2,4,7,8,8.5.

8. Graph the Richter scale rating R as a function of I/I_0.

5.4 Trigonometric Functions

In this section we will consider applications of the trigonometric functions sine, cosine, tangent, cotangent, secant, and cosecant.

Definition 5 *A function f is* periodic *if for each x in its domain $f(x+p) = f(x)$ for some positive number p. If a smallest such p exists, it is called the* period *of f.*

The frequency of a periodic function is the reciprocal of the period. The functions sine, cosine, secant, and cosecant are periodic with period 2π, and the functions tangent and cotangent are periodic with period π.

5.4.1 Problem Solver Example

A weight supported by a spring is set into motion, and the displacement of the weight from its equilibrium position after t seconds is $2\cos 4\pi t$ cm. Graph the displacement as a function of time for $0 \leq t \leq 2$, and estimate the period of this motion from the graph. The amplitude is the maximum displacement from the equilibrium position. Estimate the amplitude from the graph.

1. Start Problem Solver and select [Functions].

2. Select [Variable name] and enter `t`.

3. Select [Enter/edit functions] and select the first function.

4. Type the expression `2*cos(4*pi*t)` and then press the Enter key.

5. Press the Esc key to return to the functions menu.

6. Select [Graph functions].

7. Select [Function selection] and select the function that you entered. Press the Esc key to leave this menu.

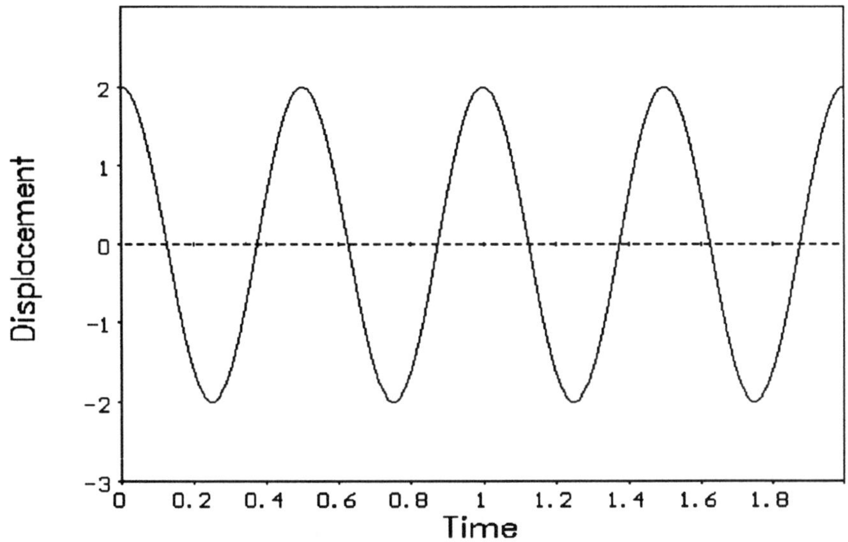

Figure 5.1: Displacement as Function of Time

8. Select [Change graph specifications] and then select the function that you entered.

9. Select [Minimum x-coordinate on axis] and enter 0.

10. Select [Maximum x-coordinate on axis] and enter 2.

11. Select [Label for x-axis] and enter `Time`.

12. Select [Label for y-axis] and enter `Displacement`.

13. Press the Esc key to leave this menu.

14. Select [Draw graph]. The graph should look like Figure 5.1 with amplitude 2, period 0.5, and frequency 2 cycles per second.

Exercise Set 5.4

1. Graph each of the following functions on the interval $[-2\pi, 2\pi]$.

   (a) $\sin 2x + \cos 3x$

   (b) $\sin^2 x + \cos^2 x$

   (c) $\tan 2x$

   (d) $\csc x + \sin x$

2. Graph each of the following functions on the interval $[-4\pi, 4\pi]$.

(a) $\sin x$

(b) $\sin 2x$

(c) $\sin 0.5x$

(d) $\sin 0.25x$

(e) Give the period for each function graphed, and guess the period of $\sin kx$.

3. Graph each of the following functions on the interval $[-2\pi, 2\pi]$.

(a) $\cot x$

(b) $\cot 2x$

(c) $\cot 0.5x$

(d) $\cot 0.25x$

(e) Give the period for each function graphed, and guess the period of $\sin kx$.

4. Use the bisection method to find all solutions of each of the following equations.

(a) $\sin x - 1 + x = 0$

(b) $\sin x = x^2$

(c) $x - \cos x = 0$

(d) $e^x + e^{-x} - 3\cos x = 0$

★ 5. Use the bisection method to find the intersections of the graphs of the equations $y = \log x$ and $y = \sin x$.

6. Suppose that a pendulum of length l feet is displaced so that the rod makes an angle of θ_0 with the vertical. If it is then released from rest, the angle θ that the rod makes with the vertical t seconds later is given by the equation

$$\theta = \theta_0 \cos \sqrt{\frac{32}{l}} t.$$

Suppose that this pendulum is 2 feet long, and that it is part of a grandfather clock.

(a) Graph θ as a function of time t for $0 \le t \le 6$.

(b) The period of a pendulum is the time it takes for the angle θ to change from θ_0 to $-\theta_0$ and back to θ_0 again. Use the graph to estimate the period of this pendulum.

(c) The amplitude of a pendulum is the maximum value of θ. Use the graph to estimate the amplitude of this pendulum.

Chapter 6

Limits

6.1 Estimating Limits

Sometimes we are interested in the behavior of $f(x)$ near a certain value of x rather than at that value of x. For example, if

$$f(x) = \frac{x^2 - 1}{x - 1}$$

$f(x)$ is not defined at $x = 1$. We may still want to know whether $f(x)$ approaches some number when x approaches but does not reach 1. We could construct a graph of the function f in order to see what happens near $x = 1$. However, in this section we will construct tables of functional values to help us solve this type of problem. Using Problem Solver to generate Table 6.1, we conclude that $f(x)$ approaches 2 as x approaches 1.

The limit of $f(x)$ as x approaches the real number a, denoted by

$$\lim_{x \to a} f(x),$$

is the real number L if $f(x)$ is near L when x is near but not equal to a. Using this notation in the example, we have

$$\lim_{x \to 1} \frac{x^2 - 1}{x - 1} = 2.$$

We may also be interested in what happens to $f(x)$ as x gets very large. The notation for this is

$$\lim_{x \to \infty} f(x) = L.$$

Consider the function

$$f(x) = \frac{3x^2 - x + 1}{2x^2 + x}.$$

x	$f(x)$
0.000	1.000
0.500	1.500
1.250	2.250
0.875	1.875
1.062	2.062
0.969	1.969
1.016	2.016
0.992	1.992
1.004	2.004
0.998	1.998

Table 6.1: $f(x) = \frac{x^2-1}{x-1}$

Table 6.2 suggests that

$$\lim_{x\to\infty} \frac{3x^2 - x + 1}{2x^2 + x} = 1.5.$$

The value of 1.5 was predictable. As x increases, x^2 increases much more rapidly. Thus for large x, $3x^2$ dominates the numerator, and $2x^2$ dominates the denominator. Consequently, as x increases, the fraction behaves more and more like $3x^2/(2x^2)$ or 1.5.

6.1.1 Problem Solver Example

We shall use Problem Solver to estimate the value of the limit

$$\lim_{x\to 1} \frac{x-1}{\log x}.$$

Note that the fraction is not defined at $x = 1$ since $\log 1 = 0$, but it is defined for values of x near 1.

1. Start Problem Solver and select [Functions].
2. Select [Enter/edit functions] and then select the first function.
3. Type the expression `(x-1)/log(x)` and press the Enter key.
4. Press the Esc key to return to the functions menu.
5. Select [Functional values] and then select the function that you just entered.
6. Select [Get x-values form formula].

x	$f(x)$
1.000	1.000
3.000	1.190
7.000	1.343
15.000	1.422
31.000	1.461
63.000	1.480
127.000	1.490
255.000	1.495
511.000	1.498
1023.000	1.499

Table 6.2: $f(x) = \frac{3x^2-x+1}{2x^2-x+1}$

7. Select Number of values and enter `10`.

8. Select $x[1]$ and enter 1.5.

9. Select $x[i]$ and enter the expression `1+(-0.5)^(i-1)`. This will create a sequence of x-values approaching 1.

10. Select Calculate x-values.

11. Select Calculate functional values. These values are shown in Table 6.3. It appears that the limit is about 2.3.

6.1.2 Problem Solver Example

We shall use Problem Solver to estimate the value of

$$\lim_{x\to\infty} \frac{x^2}{e^x}.$$

1. Start Problem Solver and select Functions.

2. Select Enter/edit functions and then select the first function.

3. Type the expression `x^2/e^x` and press the Enter key.

4. Press the Esc key to return to the functions menu.

5. Select Functional values and then select the function that you just entered.

x	$f(x)$
1.500	2.839
0.500	1.661
1.250	2.580
0.875	2.155
1.062	2.374
0.969	2.266
1.016	2.321
0.992	2.294
1.004	2.307
0.998	2.300

Table 6.3: $f(x) = \frac{x-1}{\log x}$

6. Select [Get x-values from formula].
7. Select [$x[1]$] and enter 2.
8. Select [$x[i]$] and enter the expression `x[i-1]+1`. This will create a sequence of increasing x-values.
9. Select [Calculate x-values].
10. Select [Calculate functional values]. These values are shown in Table 6.4. It appears that the limit is 0.

Exercise Set 6.1

The purpose of the first four problems is to provide practice creating a column of x-values in a function table. The particular function used for the column of y-values is not important.

1. Use Problem Solver and the function defined by $f(x) = x + 1$ to generate a function table with the following x-values. In each case, use a formula to generate the x-values.

   (a) $2^1, 2^2, 2^3, \ldots, 2^{10}$

   (b) $1 \cdot 10, 2 \cdot 10, 3 \cdot 10, \ldots, 12 \cdot 10$

   (c) $-5 \cdot 1, -5 \cdot 2, -5 \cdot 3, \ldots, -5 \cdot 20$

x	$f(x)$
2.000	0.541
3.000	0.448
4.000	0.293
5.000	0.168
6.000	0.089
7.000	0.045
8.000	0.021
9.000	0.010
10.000	0.005
11.000	0.002

Table 6.4: $f(x) = \frac{x^2}{e^x}$

(d) $-1, (-1)^2, (-1)^3, \ldots, (-1)^{-12}$

2. Use Problem Solver and the function defined by $f(x) = x - 2$ to generate a function table with the following x-values. In each case, use a formula to generate the x-values.

   (a) $1, (1/3)^1, (1/3)^2, \ldots, (1/3)^9$
   (b) $1, (-1/3)^1, (-1/3)^2, \ldots, (-1/3)^{10}$
   (c) $1, 2 - (1/3), 2 - (1/3)^2, \ldots, 2 - (1/3)^{12}$
   (d) $1, 2 - (.8), 2 - (.8)^2, \ldots, 2 - (.8)^{15}$

3. Use Problem Solver and the function defined by $f(x) = 3x - 2$ to generate a function table with the following x-values: In each case, use a formula to generate the x-values.

   (a) $1, 2, 3, \ldots, 15$
   (b) $10, 20, 30, \ldots, 100$
   (c) $5, 10, 15, 20, \ldots, 125$
   (d) $0, -1, -2, \ldots, -12$

4. Use Problem Solver and the function defined by $f(x) = x^2 + 7$ to generate a function table with the following x-values. In each case, use a formula to generate the x-values.

   (a) $1, 0.5, 0.25, 0.125, 0.0625$

(b) $1, 0.5, 0.25, 0.125, \ldots, 0.001953125$

(c) $1, 0.1, 0.01, 0.001, \ldots 0.000000001$

(d) $0, -1, -2, \ldots, -30$

5. Use Problem Solver to create function tables for each of the following limits. Either give a numerical estimate of the limit, or state that the limit does not exist.

   (a) $\lim_{x\to 0} \frac{x^2-x}{x}$

   (b) $\lim_{x\to 0} \frac{|x|}{x}$

   (c) $\lim_{x\to 0} \frac{x^2-6x+8}{x-4}$

   (d) $\lim_{x\to 0} |x|^x$

6. Use Problem Solver to create function tables for each of the following limits. Either give a numerical estimate of the limit, or state that the limit does not exist.

   (a) $\lim_{x\to 3} \frac{x^2-9}{x-3}$

   (b) $\lim_{x\to 2} \frac{x^4-16}{x-2}$

   (c) $\lim_{x\to 4} \frac{x^2-6x+8}{x-4}$

   (d) $\lim_{x\to -1} \frac{x+1}{x^2+x}$

7. Use Problem Solver to create function tables for each of the following limits. Either give a numerical estimate of the limit, or state that the limit does not exist.

   (a) $\lim_{x\to\infty} \frac{3x^2-9}{4x^2-3}$

   (b) $\lim_{x\to\infty} \frac{x^4-16}{6x^4}$

   (c) $\lim_{x\to\infty} \frac{x^2-6x+8}{x^3-4}$

   (d) $\lim_{x\to\infty} \frac{\ln x}{x}$

8. Use Problem Solver to create function tables for each of the following limits. Either give a numerical estimate of the limit, or state that the limit does not exist.

   (a) $\lim_{x\to\infty}(1+\frac{1}{x})^x$

   (b) $\lim_{x\to\infty} \frac{\sqrt{x^2+1}}{2x}$

   (c) $\lim_{x\to\infty} \sqrt{x^2+1} - x$

(d) $\lim_{x\to-\infty} \frac{x}{1-e^{-x}}$

△ 9. Use Problem Solver to create function tables for each of the following limits. Either give a numerical estimate of the limit, or state that the limit does not exist.

(a) $\lim_{x\to 0} \frac{\sin x}{x}$
(b) $\lim_{x\to 0} \frac{1-\cos x}{x}$
(c) $\lim_{x\to\infty} e^{-x}\sin x$
(d) $\lim_{x\to-\infty} \cos x$

10. Use Problem Solver to create function tables for each of the following limits. Either give a numerical estimate of the limit, or state that the limit does not exist.

(a) $\lim_{x\to 3} e^{\ln 2x}$
(b) $\lim_{x\to 0}(\log x^2)^x$
(c) $\lim_{x\to\infty} \frac{x}{\log x}$
(d) $\lim_{x\to 1} \frac{\ln x}{x}$

11. The two Mill Creek banks are fierce competitors. When the First Bank of Mill Creek started compounding the interest on savings accounts each month, the Second Bank of Mill Creek advertised that they would compound every week. The First Bank countered that they would compound daily. The Second Bank then stated that they would compound hourly, and furthermore they would better any offer made by the First Bank. The First Bank then answered with the same offer.

During these "Compounding Wars," Jimmy was shopping for a bank in which to place some of his paper-route money. He had \$135 that he wished to place in a savings account for 1 year. Both banks paid 5.5% per annum interest. First Bank offered to compound once each minute. When Jimmy took this offer to Second Bank, they offered to compound once each second. When Jimmy returned to First Bank, they stated that they would compound 10 times each second. As Jimmy went back and forth between the banks, the compounding period got shorter and shorter.

Jimmy had learned from his father that if he deposited P dollars in an account paying an interest rate of $100r\%$ per year, compounded n times each year, his money would grow to

$$A = P\left(1+\frac{r}{n}\right)^{nt}$$

after t years. Let $P = 135$, $r = 0.055$,and $t = 1$ in the above formula. Then take the limit as n goes to infinity to find the amount in the bank at the end of the year if the compounding is done *continuously*.

x	$f(x)$
2.500	3.500
3.500	6.500
3.350	6.050
2.650	3.950
3.245	5.735
2.755	4.265
3.171	5.514
2.829	4.486
3.120	5.360
2.880	4.640
3.084	5.252
2.916	4.748
3.059	5.176
2.941	4.824
3.041	5.124

Table 6.5: $f(x) = 3x - 4$

6.2 Analyzing Limits

In the last section we stated that $\lim_{x\to a} f(x) = L$ if the functional value $f(x)$ is near L whenever x is near a. This description is not precise enough to serve as a definition since it is unclear exactly what is meant by the word *near*. We could add some precision by stating that $\lim_{x\to a} f(x) = L$ if we can force $f(x)$ to be as near L as we like by taking x sufficiently near a. To express this definition symbolically, we note that $f(x)$ is near L if $|f(x) - L|$ is small, and x is near a if $|x - a|$ is small.

Definition 6 *The* limit *of $f(x)$ as x approaches the real number a is the real number L if for each positive number ϵ, there is a corresponding positive number δ such that if $0 < |x - a| < \delta$ then $|f(x) - L| < \epsilon$.*

Problem Solver uses this definition with function tables to help you estimate limits. Suppose that we need to estimate $\lim_{x\to 3} 3x - 4$. Setting $f(x) = 3x - 4$, we use Problem Solver to generate Table 6.5. Suppose we set $\epsilon = 0.5$. We see from the table that $|f(x) - 5| < 0.5$ for the last 8 lines of the table. This means that if $0 < |x - 3| < 0.120$ then $|f(x) - 5| < 0.5$. Thus if ϵ is 0.5, the corresponding δ is 0.120. This is illustrated graphically in Figure 6.1. Whenever x is between the two vertical lines with equations $x = 5 - \delta$ and $x = 5 + \delta$, $f(x)$ is between the two horizontal lines with equations $y = 3 - \epsilon$ and $y = 3 + \epsilon$.

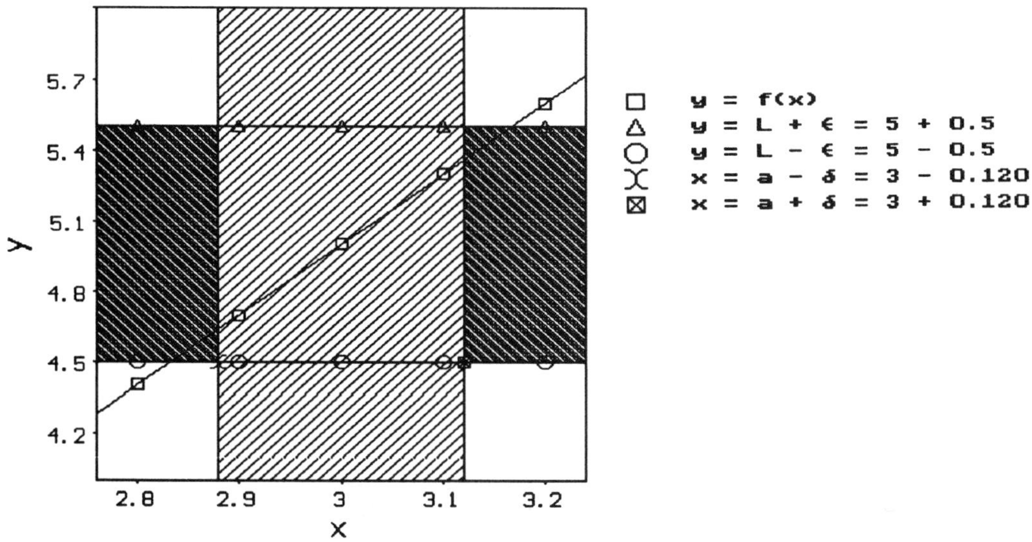

Figure 6.1: Graphical Illustration of the Limit Definition

6.2.1 Problem Solver Example

Consider

$$\lim_{x \to 0} \frac{x}{\ln(x+1)}.$$

Notice that if we try to evaluate the expression at $x = 0$ we get the meaningless expression $0/0$. We will use Problem Solver to investigate the behavior of the expression near $x = 0$.

1. Start Problem Solver and select [Functions].

2. Select [Enter/edit functions]. Then select the first function, type `x/ln(x+1)`, and press the Enter key.

3. Press the Esc key to return to the functions menu.

4. Select [Limits]. Then select the function that you just entered.

5. Select [User's estimate of the limit], select [Real number], and enter `1`.

6. Select [Calculate estimate for δ or N].

7. Select [Calculate x-values]. Look over the table of functional values to see if your guess of 1 was reasonable. Problem Solver will display a value of δ corresponding to the default value of 0.5 for ϵ.

8. Select [Graph $y = f(x)$] for a graphical illustration of the limit definition for this example.

Roughly speaking, we say that a function is continuous if we can draw its graph without lifting our pencil. That is, the graph of a continuous function has no breaks. The formal definition uses the limit concept.

Definition 7 *A function f is* continuous *at $x = a$ if all three of the following conditions are met.*

1. *The number a is in the domain of f.*

2. *$\lim_{x\to a} f(x)$ exists.*

3. *$\lim_{x\to a} f(x) = f(a)$.*

Exercise Set 6.2

For some of these problems you may need to increase the number of lines calculated in Problem Solver's table of functional values. Do this if you are quite sure that you have the correct value for the limit and if Problem Solver gives you a message indicating that your value may be incorrect. In general, as the size of ϵ is decreased you may need to increase the number of lines.

1. Consider $\lim_{x\to 5} 2x - 4$.

   (a) Use Problem Solver to help you guess the value L of the limit.

   (b) Find $\delta > 0$ such that if $0 < |x-5| < \delta$ then $|2x - 4 - L| < 0.5$.

   (c) Find $\delta > 0$ such that if $0 < |x-5| < \delta$ then $|2x - 4 - L| < 0.1$.

   (d) Find $\delta > 0$ such that if $0 < |x-5| < \delta$ then $|2x - 4 - L| < 0.01$.

2. Consider $\lim_{x\to 0} |x|$.

   (a) Use Problem Solver to help you guess the value L of the limit.

   (b) Find $\delta > 0$ such that if $0 < |x-0| < \delta$ then $||x| - L| < 0.5$.

   (c) Find $\delta > 0$ such that if $0 < |x-0| < \delta$ then $||x| - L| < 0.1$.

   (d) Find $\delta > 0$ such that if $0 < |x-0| < \delta$ then $||x| - L| < 0.01$.

3. Consider $\lim_{x\to 0}(1+2x)^{\frac{1}{x}}$. You may need to increase the number of lines in your function table to answer these questions.

   (a) Use Problem Solver to help you guess the value L of the limit.

   (b) Find $\delta > 0$ such that if $0 < |x-0| < \delta$ then $|(1+2x)^{\frac{1}{x}} - L| < 0.5$.

   (c) Find $\delta > 0$ such that if $0 < |x-0| < \delta$ then $|(1+2x)^{\frac{1}{x}} - L| < 0.1$.

(d) Find $\delta > 0$ such that if $0 < |x - 0| < \delta$ then $|(1+2x)^{\frac{1}{x}} - L| < 0.01$.

4. Consider $\lim_{x \to 0} \frac{x}{\log(x+1)}$.

   (a) Use Problem Solver to help you guess the value L of the limit.
   (b) Find $\delta > 0$ such that if $0 < |x - 0| < \delta$ then $|\frac{x}{\log(x+1)} - L| < 0.5$.
   (c) Find $\delta > 0$ such that if $0 < |x - 0| < \delta$ then $|\frac{x}{\log(x+1)} - L| < 0.1$.
   (d) Find $\delta > 0$ such that if $0 < |x - 0| < \delta$ then $|\frac{x}{\log(x+1)} - L| < 0.01$.

5. Consider $\lim_{x \to -1} x \log(2 + x)$.

   (a) Use Problem Solver to help you guess the value L of the limit.
   (b) Find $\delta > 0$ such that if $0 < |x - (-1)| < \delta$ then $|x \log(2 + x) - L| < 0.5$.
   (c) Find $\delta > 0$ such that if $0 < |x - (-1)| < \delta$ then $|x \log(2 + x) - L| < 0.1$.
   (d) Find $\delta > 0$ such that if $0 < |x - (-1)| < \delta$ then $|x \log(2 + x) - L| < 0.01$.

6. Consider $\lim_{x \to 0} (x^2)^x$.

   (a) Use Problem Solver to help you guess the value L of the limit.
   (b) Find $\delta > 0$ such that if $0 < |x - 0| < \delta$ then $|(x^2)^x - L| < 0.5$.
   (c) Find $\delta > 0$ such that if $0 < |x - 0| < \delta$ then $|(x^2)^x - L| < 0.1$.
   (d) Find $\delta > 0$ such that if $0 < |x - 0| < \delta$ then $|(x^2)^x - L| < 0.01$.

7. Let L $= \lim_{x \to 0} 4x^3 - x$.

   (a) Estimate L.
   (b) Let $\epsilon = 0.001$, and use Problem Solver to find a corresponding δ such that if $0 < |x - 0| < \delta$ then $|4x^3 - x - L| < \epsilon$.

8. Use Problem Solver to construct a graph illustrating the limit definition. Discuss this graph.

△ 9. Consider $\lim_{x \to 0} \frac{\sin x}{x}$.

   (a) Use Problem Solver to help you guess the value L of the limit.
   (b) Find $\delta > 0$ such that if $0 < |x - 5| < \delta$ then

$$|\frac{\sin x}{x} - L| < 0.5.$$

   (c) Find $\delta > 0$ such that if $0 < |x - 5| < \delta$ then

$$|\frac{\sin x}{x} - L| < 0.1.$$

(d) Find $\delta > 0$ such that if $0 < |x - 5| < \delta$ then

$$|\frac{\sin x}{x} - L| < 0.01.$$

10. Let

$$\lim_{x \to 2} \frac{x^2 - 4}{x - 2} = L.$$

(a) Estimate L.

(b) Let $\epsilon = 0.5$, and use Problem Solver to find a corresponding δ such that if $0 < |x - 2| < \delta$ then $|\frac{x^2-4}{x-2} - L| < \epsilon$.

(c) Use Problem Solver to construct a graph illustrating the limit definition. Discuss this graph.

⋆ 11. Use Problem Solver to decide whether each of the following functions is continuous at $x = a$. First check to see that $f(a)$ is defined. If $f(a)$ exists, go on to the second step. Check to see whether or not $\lim_{x \to a} f(x) = f(a)$.

(a) $f(x) = \begin{cases} \frac{x^2-9}{x+3} & \text{if } x \neq -3 \\ -6 & \text{if } x = -3 \end{cases}$

$a = -3$.

(b) $f(x) = [\![x]\!]$, $a = 2$.

(c) $f(x) = \begin{cases} |x|^x & \text{if } x \neq 0 \\ 0 & \text{if } x = 0 \end{cases}$

$a = 0$.

(d) $f(x) = \begin{cases} \frac{x-1}{\ln x} & \text{if } x > 0 \text{ and } x \neq 1 \\ 1 & \text{if } x = 1 \end{cases}$

$a = 1$.

Chapter 7

Derivatives

7.1 Estimating Derivatives

A function f defines a relationship between an independent variable x and a dependent variable y, which we denote by $f(x)$. For example, consider the function f given in Table 7.1. The functional value $f(x)$ gives the distance in feet from the starting position that a certain van can travel in x seconds. This function relates distance to time. The distance traveled during the time interval beginning 5 seconds after the start and ending 8 seconds after the start is given by $f(8) - f(5) = 301.960 - 141.760 = 160.2$ feet. The average velocity during this time interval is

$$\frac{f(8) - f(5)}{8 - 5} = \frac{160.2}{3} = 53.4 \text{ ft/sec.}$$

This quotient gives the average rate of change of distance with respect to time. Rates of change are extremely important in the study of changing quantities. If x changes from a to $a + h$, then $f(x)$ changes from $f(a)$ to $f(a+h)$. Thus the change in $f(x)$ is $f(a+h) - f(a)$. The average rate of change of $f(x)$ with respect to x is

$$\frac{f(a+h) - f(a)}{h}.$$

Again consider the data in Table 7.1. Suppose that we are interested in the velocity of the van 2 seconds after the start. We can approximate this using average velocities. The average velocity during the following 4 seconds, that is, for $x \in [2, 6]$, is

$$\frac{f(2+4) - f(2)}{4} = \frac{f(6) - f(2)}{4} = \frac{189.546 - 35.974}{4} = 38.393 \text{ ft/sec.}$$

Since the van is accelerating, this estimate is higher than the actual speed 2 seconds after the start. To get a more accurate estimate, we should consider a shorter time

x	$f(x)$
0	00.000
1	14.316
2	35.974
3	64.650
4	100.020
5	141.760
6	189.546
7	243.054
8	301.960
9	365.940
10	434.670

Table 7.1: Distance $f(x)$ Traveled in x Seconds

interval. The average velocity during the following 2 seconds, that is, for $x \in [2, 4]$, is

$$\frac{f(2+2)-f(2)}{2} = \frac{f(4)-f(2)}{2} = \frac{100.020-35.970}{2} = 32.023 \text{ ft/sec}.$$

To get an even more accurate estimate, consider average velocity during the following second, that is, for $x \in [2, 3]$. This average velocity is

$$\frac{f(2+1)-f(2)}{1} = f(3)-f(2) = 64.650-35.970 = 28.68 \text{ ft/sec}.$$

It looks as though we need to use an even shorter time interval. We could do this if we had an equation giving the distance $f(x)$ that the van travels in x seconds. Suppose that the function f is given by the equation

$$f(x) = 0.054x^3 + 3.833x^2 + 10.537x.$$

The average velocity during the time interval $[2, 2+h]$ is given by

$$\frac{f(2+h)-f(2)}{h}.$$

From Table 7.1, it appears that the average velocity approaches approximately 25 ft/sec as h approaches 0. Using limit notation we have

$$\lim_{h\to 0} \frac{f(2+h)-f(2)}{h} \approx 25.$$

We shall use this idea to define *instantaneous* velocity.

h	$\frac{f(2+h)-f(2)}{2}$
4.000	38.393
2.000	32.023
1.000	28.676
0.500	26.962
0.250	26.095
0.125	25.659
0.062	25.440
0.031	25.331
0.016	25.276
0.008	25.248

Table 7.2: Average Velocity between 2 and $2 + h$ Seconds

Definition 8 *If $f(x)$ gives the distance that an object travels in x seconds, then the* instantaneous velocity *of that object when $x = a$ is given by*

$$\lim_{h \to 0} \frac{f(a+h) - f(a)}{h}$$

provided that this limit exists.

For the van, the instantaneous velocity is the speedometer reading. The limit used to define instantaneous velocity is used in other settings as well. In fact, it comes up often enough that it is given a special name.

Definition 9 *The* derivative *of a function f at $x = a$, denoted by $f'(a)$, is given by*

$$f'(a) = \lim_{h \to 0} \frac{f(a+h) - f(a)}{h}$$

provided that this limit exists. By replacing a with the variable x, we get the derivative function $f'(x)$.

7.1.1 Problem Solver Example

Let $f(x) = x^2$. We shall estimate $f'(3)$ by examining the quotient

$$\frac{f(3+h) - f(3)}{h} = \frac{(3+h)^2 - 3^2}{h}$$

as h approaches 0.

1. Start Problem Solver and select Functions.

2. Select Variable name and enter `h`.

3. Select Enter/edit functions and then select the first function.

4. Type the expression `((3+h)^2-3^2)/h` and press the Enter key. Then press the Esc key to return to the functions menu.

5. Select Functional values and then select the function that you just entered.

6. Select Get x-values from formula.

7. Select $x[1]$ and enter 1.

8. Select $x[i]$ and enter `- 0.5*x[i-1]`.

9. Select Calculate x-values.

10. Select Calculate functional values.

The functional values should be approaching 6. Problem Solver provides a more direct way of approximating derivatives. We shall illustrate this second method in the next example.

7.1.2 Problem Solver Example

Again let $f(x) = x^2$. We shall construct a table of values of the derivative $f'(x)$ for $x \in \{1, 1.1, 1.2, \ldots, 4.0\}$.

1. Start Problem Solver and select Functions.

2. If the variable name is not x select Variable name, and enter `x`.

3. Select Enter/edit functions, select the first function, and then enter `x^2`.

4. Press the Esc key to return to the functions menu.

5. Select Derivatives and then select the function that you just entered.

6. Select Compute table.

7. Select Minimum x-value and enter `1`.

8. Select Maximum x-value and enter `4`.

9. Select Number of x-values. Since we want $(4-1)/0.1 = 30$ steps or 31 points, enter `31`.

10. Select [Generate table].

11. Select [Calculate x-values]. Use the PgDn key to move down the table, and observe that $f'(3) = 6$.

Suppose that an object is moving along a line and that its position at time t is given by $s(t)$. The velocity $v(t) = s'(t)$ is the rate of change of the position with respect to time. The acceleration $a(t)$ is the rate of change of the velocity with respect to time. Thus $a(t) = v'(t)$. Notice that the acceleration is the derivative of the derivative of the position. We say that the acceleration is the second derivative of the position, and we write $a(t) = s''(t)$.

Theorem 6 *If* $f(x) = a_n x^n + a_{n-1}x^{n-1} + \cdots + a_1 x + a_0$, *then* $f'(x) = na_n x^{n-1} + (n-1)a_{n-1}x^{n-2} + \cdots + a_1$.

This theorem makes it easy to find derivatives of polynomials. For example, if $f(x) = 4x^3 - 3x^2 + 8$, then $f'(x) = 3 \cdot 4x^{3-1} + 2 \cdot (-3)x^{2-1} = 12x^2 - 6x$.

Exercise Set 7.1

1. Let f be defined by the following table:

x	$f(x)$
1	2.00
2	2.50
3	2.75
4	3.00
5	3.10

Find the average rate of change of $f(x)$ as x changes from

(a) 2 to 4.

(b) 1 to 5.

(c) 1 to 4.

(d) 1 to 3.

(e) 1 to 2.

2. The following table consists of data taken from an automobile test.

Time (seconds)	Distance Traveled (feet)
1	15
2	38
3	70
4	120

Find the average velocity of the automobile over the time interval

(a) $[2,4]$

(b) $[1,4]$

(c) $[1,3]$

(d) $[1,2]$

3. For each of the following functions, estimate $f'(3)$ by calculating

$$\frac{f(3+h)-f(3)}{h}$$

for $h=(-0.5)^i$ for $i=0,1,\ldots,10$.

(a) $f(x)=5$

(b) $f(x)=4x-2$

(c) $f(x)=3x^2-5x$

(d) $f(x)=\sqrt{x}$

(e) $f(x)=1/x$

(f) $f(x)=\log x$

△ 4. For each of the following functions, estimate $f'(0)$ by calculating

$$\frac{f(0+h)-f(0)}{h}$$

for $h=(-0.5)^i$ for $i=0,1,\ldots,10$.

(a) $f(x)=\sin x$

(b) $f(x)=\cos x$

(c) $f(x)=\tan x$

(d) $f(x)=\sec x$

(e) $f(x)=\sin x\cos x$

(f) $f(x)=\sin^2 x+\cos^2 x$

5. For each of the following functions, use Problem Solver to generate a table of values of the first and second derivatives for $x\in\{1,1.2,1.4,\ldots,5\}$.

(a) $f(x)=5$

(b) $f(x)=3x-2$

(c) $f(x)=-4x+3$

(d) $f(x) = 2x^2$

(e) $f(x) = e^x$

(f) $f(x) = \log x$

(g) $f(x) = 1/x$

△ 6. For each of the following functions, use Problem Solver to generate a table of values of the first and second derivatives for $x \in \{0, 0.2, 0.4, \ldots, 4\}$.

(a) $f(x) = \sin x$

(b) $f(x) = \cos x$

(c) $f(x) = \sin x + \cos x$

7.2 Tangents and Solving Equations

7.2.1 Tangent Lines

Consider the line tangent to the circle with equation $x^2 + y^2 = 25$ at the point $(-2, \sqrt{21})$. From high school geometry, we know that any line intersecting a circle at a single point is called a tangent line. The limitations of this definition become readily apparent if we consider the following problem: find the line tangent to the *top half* of the circle with equation $x^2 + y^2 = 25$ at the point $(-2, \sqrt{21})$. Surely this tangent line must be the same as the tangent to the whole circle, yet other lines, such as the vertical line with equation $x = -2$, intersect the half circle at a single point. We need to reformulate the definition of *tangent line* so that it will apply to any curve defined by an equation of the form $y = f(x)$. Solving the equation of the circle for y, we find that the upper half of the circle is the graph of $f(x) = \sqrt{25 - x^2}$. The following table lists values of this function.

x	$f(x) = \sqrt{25 - x^2}$
-2	4.583
-1	4.899
0	5.000
1	4.899
2	4.583

The secant line passing through the points $(-2, 4.583)$ and $(2, 4.583)$ shown in Figure 7.1 could be considered an approximation to the tangent line. It has slope $(4.583 - 4.583)/(2 - (-2)) = 0$. The secant line passing through $(-2, 4.583)$ and $(1, 4.899)$ could be considered a better approximation. It has slope $(4.899 - 4.583)/(1 - (-2)) = 0.105$. The best approximation for the tangent that is available

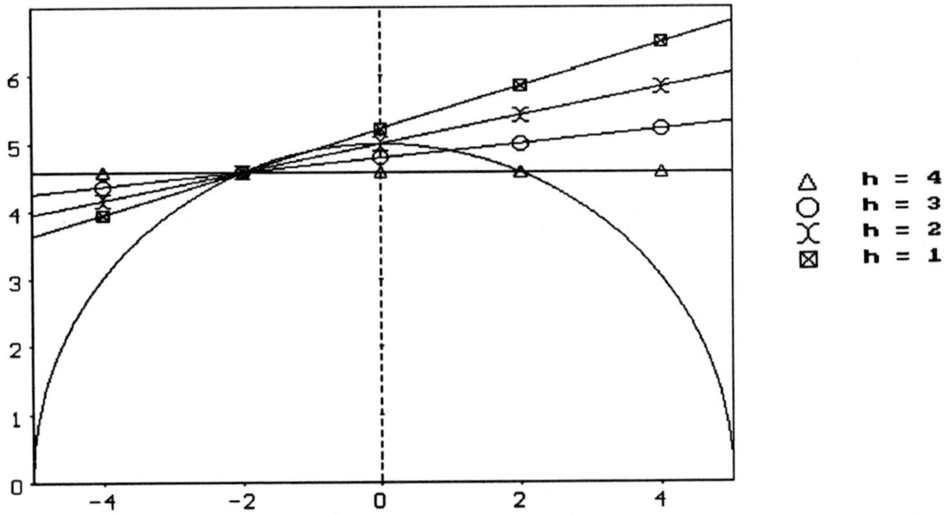

Figure 7.1: Approximations to the Tangent Line

from the table is the secant line passing through (-2,4.583) and (-1,4.899). It has slope

$$\frac{f(-1)-f(-2)}{1} = 4.899 - 4.583 = 0.316.$$

The slopes of these approximations to the tangent line are of the form

$$\frac{f(-2+h)-f(-2)}{h},$$

and these approximations get better as h approaches 0. From Table 7.3, it appears that

$$\lim_{h\to 0}\frac{f(-2+h)-f(-2)}{h} \approx 0.436.$$

We recognize this limit as $f'(-2)$. The line passing through $(-2, \sqrt{21})$ with slope 0.436 is called the tangent line.

Definition 10 *The line* tangent *to the graph of* $y = f(x)$ *at the point* $(a, f(a))$ *is the line passing through the point* $(a, f(a))$ *with slope* $f'(a)$ *provided the derivative of* f *exists at* $x = a$.

7.2.2 Problem Solver Example

We shall use Problem Solver to find the equation of the line tangent to the graph of $y = \log x$ at the point (1,0). We need the value of the derivative of $\log x$ at $x = 1$. The easiest way to do this is to let Problem Solver generate a table of derivative values for x near 1.

h	$\frac{f(-2+h)-f(-2)}{h}$
-0.5000	0.5049
0.2500	0.4047
-0.1250	0.4529
0.0625	0.4284
-0.0313	0.4234
0.0156	0.4344
-0.0078	0.4375
0.0039	0.4359
-0.0020	0.4367
0.0010	0.4363
-0.0005	0.4365
0.0002	0.4364
- 0.0001	0.4365

Table 7.3: Slope of Tangent Line as Limit

1. Start Problem Solver and select [Functions].
2. Select [Enter/edit functions] and then select the first function.
3. Type `log(x)` and press the Enter key.
4. Press the Esc key to return to the functions menu.
5. Select [Derivatives] and then select the function that you entered.
6. Select [Compute table].
7. Select [Minimum x-value] and enter 0.99.
8. Select [Maximum x-value] and enter 1.01.
9. Select [Number of x-values] and enter 5.
10. Select [Generate table].
11. Select [Calculate x-values].

Reading the output, we see that the derivative of $\log x$ at $x = 1$ is approximately 0.434. Using the point-slope form of the equation of a line, we find that

$$y - 0 = 0.434(x - 1)$$

or

$$y = 0.434x - 0.434$$

is the equation of the tangent line.

7.2.3 Problem Solver Example

Problem Solver can be used to graph a sequence of secant lines approaching the line tangent to a curve at a given point. Consider the function given by $f(x) = x \log x$ at $x = 0.5$. We will graph the lines passing through $(0.5, f(0.5)$ and each of the points $(x_i, f(x_i))$ where $x_i = 0.5 + 4(0.7)^{i-1}$ for $i \in 0, 1, 2, 3, 4$.

1. Start Problem Solver and select [Functions].
2. Select [Enter/edit functions], type `x*log(x)`, and press the Enter key.
3. Press the Esc key to return to the functions menu.
4. Select [Graph functions].
5. Select [Change graph specifications] and then select the function that you entered.
6. Select [Minimum x-coordinate on axis] and enter `0`.
7. Select [Maximum x-coordinate on axis] and enter `3`.
8. Select [Minimum y-coordinate on axis] and enter `-1`.
9. Select [Maximum y-coordinate on axis] and enter `2`.
10. Select [Derivative] and then select the function that you entered.
11. Select [Explore secant and tangent lines].
12. Select [Approximate tangent at x] and enter `0.5` for the value of x.
13. Select [Beginning step size] and enter `2`.
14. Select [Calculate values].
15. From the next menu, again select [Calculate values].
16. When the graph of the function appears, press the Enter key to add secant lines to the graph.

7.2.4 Using the Derivative to Solve Equations

Consider the problem of solving an equation of the form $f(x) = 0$. As an example, let $f(x) = x^5 - x - 1$ so that we are solving the equation $x^5 - x - 1 = 0$. A moment of reflection will convince you that the methods learned in high school such as factoring and the quadratic formula will not help in this problem. In fact, there is no method for finding exact solutions. This means that we must settle for an approximation to the solution (or solutions). We shall replace the function with a simpler function and then solve the new problem. This is a common technique for approximating solutions of difficult problems. If we replace the function with a linear function of the form $mx + b$, we will then be solving the equation

$$mx + b = 0,$$

which has solution $x = -b/m$. If we replace the graph of f with a line, perhaps the best line to use is the tangent line at some point $(a, f(a))$. In doing this, we are making the assumption that the graph of f is almost a line if we restrict our attention to values of x very near a. Consequently, we can see that the choice of a is critical. We will choose a to be our best guess for the solution we seek. This guess could come from graphing the function and guessing where it crosses the x-axis. For our example, we will choose $a = 1$. Since $f'(x) = 5x^4 - 1$, the line tangent to the graph at $x = a = 1$ has slope $f'(a) = f'(1) = 5 - 1 = 4$. Also the tangent passes through the point $(a, f(a)) = (1, f(1)) = (1, -1)$. This means that the equation of the tangent line is

$$y - f(a) = f'(a)(x - a)$$

or

$$y - (-1) = 4(x - 1).$$

To find where this tangent line crosses the x-axis, we set $y = 0$ and solve for x. This yields

$$x = a - \frac{f(a)}{f'(a)}$$

or, in the case of our example,

$$x = 1 - \frac{-1}{4} = 1.25.$$

We now have a new estimate for the solution of our equation, namely 1.25. We begin again setting $a = 1.25$.

The method introduced above is called *Newton's method.* To describe this algorithm more generally, denote our initial guess by x_0. We then set

$$x_1 = x_0 - \frac{f(x_0)}{f'(x_0)},$$

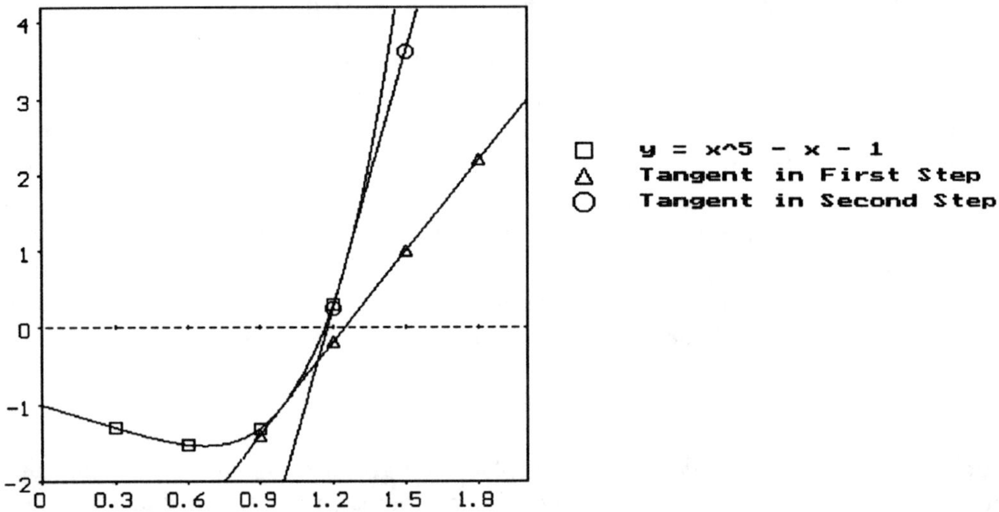

Figure 7.2: Newton's Method

$$x_2 = x_1 - \frac{f(x_1)}{f'(x_1)},$$

and

$$x_3 = x_2 - \frac{f(x_2)}{f'(x_2)}.$$

We continue until the x_is appear to be approaching some number. With a good first guess, this usually takes only a few steps. See Figure 7.2 for a graphical representation of Newton's method. Newton's method is usually expressed in the following form: Let x_0 be an initial guess for the solution of $f(x) = 0$. Then for $n = 0, 1, \ldots$, set

$$x_{n+1} = x_n - \frac{f(x_n)}{f'(x_n)}.$$

7.2.5 Problem Solver Example

We shall use Newton's method to find all solutions of the equation $x^2 - 3x = 6$. To apply Newton's method to this equation, we need to express it in the form $f(x) = 0$. Since the equation is equivalent to $x^2 - 3x - 6 = 0$, we set $f(x) = x^2 - 3x - 6$.

1. Start Problem Solver and select [Functions].

2. Select [Enter/edit functions] and then select the first function.

3. Type `x^2-3*x-6` and press the Enter key.

4. Press the Esc key to return to the functions menu.

5. To find a good initial guess, we shall first graph the function. Select [Graph functions].

6. Select [Function selection] and select the function that you entered. Press the Esc key to leave this menu.

7. Select [Draw graph]. Notice that the graph crosses the x-axis near $x = -1$ and near $x = 4$.

8. To get a better view of the graph, select [Change graph specifications].

9. Select the function that you entered.

10. Select [Minimum x-value to graph] and enter -10.

11. Select [Maximum x-value to graph] and enter 10.

12. Select [Minimum y-coordinate on axis] and enter -50.

13. Select [Maximum y-coordinate on axis] and enter 50.

14. Press the Esc key to leave this menu.

15. Select [Draw graph].

16. Press the Esc key twice to return to the functions menu.

17. Select [Solve $f(x) = 0$] and then select the function that you entered.

18. Select [Method] and then select [Newton's method].

19. Select [Derivative] and enter `2*x-3`.

20. Select [Initial guess] and enter -1.

21. Select [Options] and then select [Intermediate calculations displayed]. Request that intermediate calculations be displayed.

22. Press the Esc key to leave the options menu.

23. Select [Compute solution] and observe that the first solution is approximately -1.37228.

24. Select [Initial guess] and enter 4.

25. Select [Compute solution] and observe that the second solution is approximately 4.37228.

You may also specify the maximum number of steps to use, and the largest allowable error and residual error. If x is an estimate for x^* where x^* is a solution of $f(x) = 0$, the error is $|x - x^*|$, and the residual error is $|f(x)|$. Since x^* is not known, Problem Solver uses an estimate of the error. Problem Solver will keep computing new approximate solutions until either the maximum number of iterations has been reached or all of the error tolerances have been met.

With a good first guess, Newton's method usually produces accurate results in just a few steps. In many cases the number of accurate decimal places doubles in each step. This has made Newton's method one of the most popular methods for solving equations. However, the method has two shortcomings. If the portion of the graph of f near the solution of $f(x) = 0$ and your initial guess varies too much from a line, the method can fail. This means that it is very important to use good first guesses. A second weakness is that we are required to compute a derivative.

We can address the second weakness of Newton's method. From the definition of the derivative, we know that

$$f'(x_n) = \lim_{h \to 0} \frac{f(x_n + h) - f(x_n)}{h}.$$

Thus for h near 0

$$f'(x_n) \approx \frac{f(x_n + h) - f(x_n)}{h}.$$

Setting $h = x_{n-1} - x_n$, we have

$$f'(x_n) \approx \frac{f(x_{n-1}) - f(x_n)}{x_{n-1} - x_n}.$$

In Newton's method

$$x_{n+1} = x_n - \frac{f(x_n)}{f'(x_n)}.$$

Using our approximation for the derivative, we get

$$x_{n+1} = x_n - \frac{f(x_n)}{\frac{f(x_{n-1}) - f(x_n)}{x_{n-1} - x_n}},$$

which simplifies to

$$x_{n+1} = x_n - \frac{f(x_n)(x_{n-1} - x_n)}{f(x_{n-1}) - f(x_n)}.$$

Since x_{n+1} is defined in terms of x_n and x_{n-1}, two initial guesses are required.

This modification of Newton's method is called the *secant method.* To use this algorithm, we let x_0 and x_1 be two initial guesses for the solution of $f(x) = 0$. Then for $n \geq 1$, we set

$$x_{n+1} = x_n - \frac{f(x_n)(x_{n-1} - x_n)}{f(x_{n-1}) - f(x_n)}.$$

7.2.6 Problem Solver Example

We shall use the secant method to find the solution of the equation $x^2 - 3x = 6$ near $x = 4$.

1. Start Problem Solver and select Functions.
2. Select Enter/edit functions and then select the first function.
3. Type `x^2-3*x-6` and press the Enter key.
4. Press the Esc key to return to the functions menu.
5. Select Solve $f(x) = 0$ and then select the function that you entered.
6. Select Method and then select Secant method.
7. Select Initial guess. Two initial guesses will be requested. Enter 3 and 4.
8. Select Compute solution and observe that the solution is approximately 4.37228.

Exercise Set 7.2

1. Use Problem Solver to estimate the slope of the line tangent to the graph of f at the indicated x-value.
   (a) $f(x) = 3x - 2,\ x = 4$
   (b) $f(x) = e^{2x},\ x = 0$
   (c) $f(x) = x \log x,\ x = 2$
   (d) $f(x) = 4x^3 - 3x^2,\ x = 1$
2. Use Problem Solver to estimate derivatives, and then use the point-slope form of the equation of a line to find the equation of the line tangent to each of the following functions at the indicated value of x.
   (a) $f(x) = -3x - 8,\ x = 4$
   (b) $f(x) = 4,\ x = 2$
   (c) $f(x) = \frac{x^2}{x^2+1},\ x = 0$
   (d) $f(x) = \frac{x^2}{x^2+1},\ x = 1$
   (e) $f(x) = x^{-1} \log x + x^{-2},\ x = 10$
3. Use Newton's method to find all solutions of each of the following equations:

(a) $3x - 9 = 0$

(b) $x^3 - 7x = 10$

(c) $x^2 - 1 = 0$

(d) $x^4 - x^3 - x^2 = 0$

(e) $4x^3 - 3x = 0$

4. Use the secant method to find all solutions of each of the following equations:

(a) $4x - 6 = 0$

(b) $x^2 - 4 = 0$

(c) $x - \log x = 2$

(d) $4 - x^2 = 2^x$

(e) $\frac{x^3-1}{x^2+3x+9} = 1$

△ 5. Use the secant method to find all solutions of each of the following equations:

(a) $\sin x = x - 1$

(b) $\sin x + \cos x - x = 0$

(c) $x - \quad cosx = 0$

(d) $x^4 - \cos x = 0$

6. Consider each of the following equations:

$$\begin{aligned} x^3 + x^2 + x + 2 &= 0 \\ x^3 - 12x^2 + 1 &= 0 \end{aligned}$$

(a) Find all solutions of both equations using Newton's method.

(b) Find all solutions of both equations using the secant method.

(c) Find all solutions of both equations using the bisection method.

(d) Discuss the relative advantages and disadvantages of the three methods. Include in your discussion comments about the number of steps required for each of the methods. Also state which method you prefer, and give reasons for your choice.

★★ △ 7. Again consider the bridge described in Exercise 12 of Section 3.1. This time use a polynomial of degree 3 to determine the shape of the bridge. The connecting points between the bridge and the road should be perfectly smooth now. Graph the bridge and the two approaches on the same coordinate system. *Hint:* The slope of the east approach is $\tan 7°$.

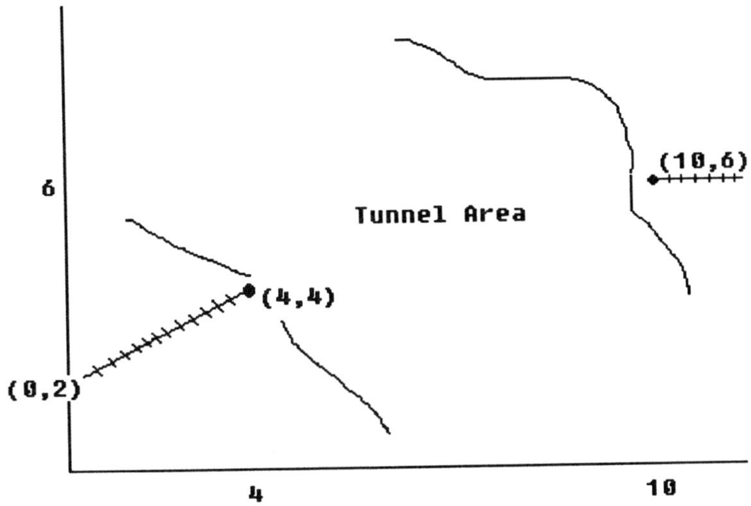

Figure 7.3: Railroad Project

★★△ 8. An airplane is cruising at 10,000 feet 10 miles from the runway. Find a third degree polynomial whose graph makes an appropriate flight path for this plane as it comes in for a landing. Graph your flight path. *Hint:* What is the slope of this path while cruising and while landing?

★★△ 9. The two railroads shown in Figure 7.3 are to be connected by digging a tunnel through the intervening mountain. Find an appropriate shape for this railroad using a polynomial of degree 3. Graph the two existing railroads and the connecting link.

7.3 Extrema

In this section we will be studying extreme values or maximum and minimum values of functions.

Definition 11 *Let the function f be defined on some domain D. If $x_M \in D$, and if $f(x_M) \geq f(x)$ for all $x \in D$, then we say that f has a* maximum value *at x_M. This maximum value is $f(x_M)$. If $x_m \in D$, and if $f(x_m) \leq f(x)$ for all $x \in D$, then we say that f has a* minimum value *at x_m. This minimum value is $f(x_m)$.*

The function f graphed in Figure 7.3 has maximum $f(-2) = f(2) = 9$. The value $f(0) = 1$, though not a maximum, would become a maximum if we reduced the size of the domain D. That is, if we restrict our attention to values of x near 0, then $f(0) = 1$ does become a maximum.

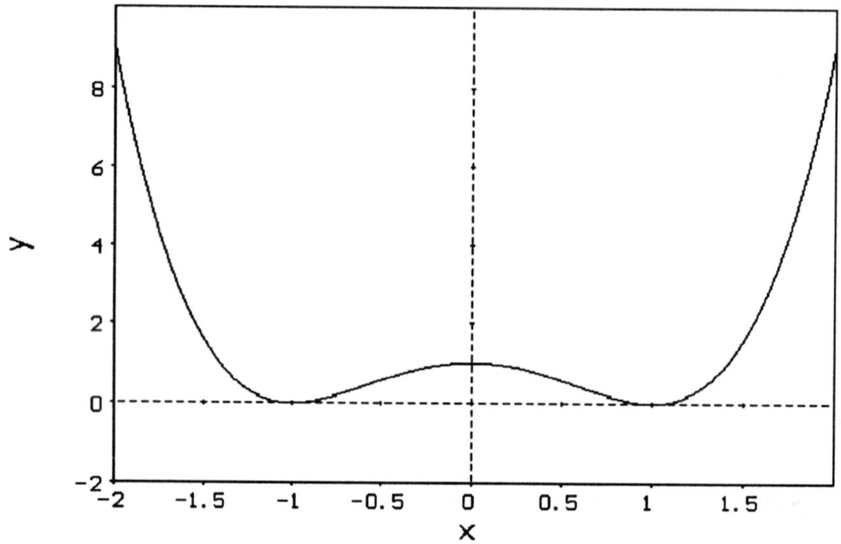

Figure 7.4: $f(x) = x^4 - 2x^2 + 1$

Definition 12 *Let the function f be defined on some domain D containing the value c. If an interval (a, b) containing c exists where $f(c) \geq f(x)$ for all $x \in (a, b)$, we say that f has a* relative *or* local maximum *at $x = c$. If $f(c) \leq f(x)$ for all $x \in (a, b)$, we say that f has a* relative *or* local minimum *at $x = c$.*

Definition 13 *We say that a function f is* increasing *on the interval $[a, b]$ if for each x_1 and x_2 in $[a, b]$, $x_1 < x_2$ implies that $f(x_1) < f(x_2)$. We say that f is* decreasing *on $[a, b]$ if for each x_1 and x_2 in $[a, b]$, $x_1 < x_2$ implies that $f(x_1) > f(x_2)$.*

This definition uses a left-to-right orientation. To see this, imagine that you are walking along the graph of f from left to right (in the direction of increasing x). If you are walking uphill, the function is increasing. On the other hand, if you are walking downhill, the graph is decreasing. The function graphed in Figure 7.3 is increasing on the intervals $[-1, 0]$ and $[1, 2]$ and decreasing on the intervals $[-2, -1]$ and $[0, 1]$. It is not surprising that a function is increasing if its tangent lines have positive slope (sloping upward), and decreasing if its tangent lines have negative slope (sloping downward).

Theorem 7 *If f is continuous on $[a, b]$, and if $f'(x) > 0$ for all $x \in (a, b)$, then f is increasing on $[a, b]$, and if $f'(x) < 0$ for all $x \in (a, b)$, then f is decreasing on $[a, b]$.*

This leads to the following theorem, which is known as the *First Derivative Test:*

Theorem 8 *Let f be a function defined at $x = c$. Suppose that an interval (a, b) containing c exists where $f(x)$ is defined for each $x \in (a, b)$.*

1. *If $f'(x) < 0$ for $x \in (a, c)$ and $f'(x) > 0$ for $x \in (c, b)$, then f has a local minimum at $x = c$.*

2. *If $f'(x) > 0$ for $x \in (a, c)$ and $f'(x) < 0$ for $x \in (c, b)$, then f has a local maximum at $x = c$.*

7.3.1 Problem Solver Example

We shall use Problem Solver to use the First Derivative Test to identify all local extrema (minima or maxima) $f(x) = 3x^3 - 9x + 1$ for $x \in [-2, 3]$. In addition, we will use Problem Solver to give the minimum and maximum values of f for $x \in [-2, 3]$.

1. Start Problem Solver and select [Functions].

2. Select the first function, type `3*x^3-9*x+1`, and press the Enter key.

3. Press the Esc key to return to the functions menu.

4. Select [Derivatives] and then select the function that you just entered.

5. Select [Compute table].

6. Select [Minimum x-value] and enter -2.

7. Select [Maximum x-value] and enter 3.

8. Select [Number of x-values] and enter 51.

9. Select [Generate table].

10. Select [Calculate x-values].

11. Use the PgUp and PgDn keys to view different portions of the table. Note that for $x < -1$, the derivative is positive. Note also that the values of $f(x)$ are increasing. For $x \in (-1, 1)$ the derivative is negative, and the values of $f(x)$ are decreasing. Thus $f(-1) = 7$ is a local maximum. For $x > 1$ the derivative is again positive, and the values of $f(x)$ are increasing. This makes $f(1) = -5$ a local minimum. We still do not know whether 5 is the minimum value of $f(x)$ or whether 7 is the maximum value of $f(x)$ for $x \in [-2, 3]$. It is possible that $f(-2)$ or $f(3)$ could be the maximum or minimum.

12. Press the Esc key twice to return to the functions menu.

13. Select [Maxima/minima] and then select the function that you entered.

14. Select [Left end-point a] and enter -2.

15. Select | Right end-point b | and enter 3.

16. Select | Compute extrema on $[a, b]$ |. The minimum is $f(-2) = -5$ (the same as the local minimum), and the maximum is $f(3) = 55$.

Exercise Set 7.3

1. Graph each of the following functions, and then use the graph to estimate the intervals where the function is increasing and the intervals where the function is decreasing.

   (a) $f(x) = x^2 - 3x + 2$

   (b) $f(x) = |\log x|$

   (c) $f(x) = x^3 - 12x + 1$

   (d) $f(x) = \frac{x^2}{x^2+1}$

2. For each of the following functions, graph both the function and its derivative. Then estimate the intervals where the function is increasing and where the function is decreasing. Finally, estimate the intervals where the derivative is positive and the intervals where the derivative is negative.

   (a) $f(x) = 1 - x^2$

   (b) $f(x) = x^3 - 3x$

   (c) $f(x) = x - \log x$

   (d) $f(x) = 2^x + 2^{-x}$

3. For each of the following functions, create a table of n values of the derivative over the given interval. Using this derivative table, estimate the intervals where the function is increasing and those where the function is decreasing.

   (a) $f(x) = x^2 - 6x + 9$ for $x \in [-5, 5]$, $n = 11$

   (b) $f(x) = x^3 - x^2$ for $x \in [-1, 3]$, $n = 41$

   (c) $f(x) = \frac{2x^4+4x^3-1}{x^4+1}$ for $x \in [-10, 10]$, $n = 201$

   (d) $f(x) = x^4 - e^x$ for $x \in [0, 10]$, $n = 101$

4. Find the minimum and maximum values over the given interval for each of the functions given in Exercise 3.

5. Find the minimum and maximum values for each of the following functions on the given interval.

(a) $f(x) = 4x^3 - 3x$ for $x \in [-1, 2]$
(b) $f(x) = 2x^2 - 1$ for $x \in [-2, 0]$
(c) $f(x) = 2 + 3x + x^2 - x^8$ for $x \in [0, 2]$
(d) $f(x) = 4 + x^2 - 2^x$ for $x \in [-5, 5]$

△ 6. Find the minimum and maximum values for each of the following functions on the given interval.

(a) $f(x) = \sin x + \cos x$ for $x \in [-\pi, \pi]$
(b) $f(x) = \tan^2 x + \sin x$ for $x \in [0, \pi/4]$
(c) $f(x) = \sin^2 x + \cos^2 x$ for $x \in [-\pi, \pi]$
(d) $f(x) = \sec x + \cos x$ for $x \in [-\pi/4, \pi/4]$

7. When Jimmy Greene was a small child, Mr. Greene built a rectangular play area for him. Being an expert at parenting, Mr. Greene wanted to make the play area as large as possible. If he had 89 feet of fencing material, find the dimensions of the play area that would allow the greatest possible area for little Jimmy.

8. The Mill Creek Motor Lodge charges $25.50 per night for its rooms. Ordinarily 75% of its 24 rooms are rented. It is estimated that for each $5 increase in the price their occupancy will go down by 20%. It is also assumed that for each $5 decrease in the price, their occupancy would go up by 20% until all the rooms are used. Find the price that will maximize the total income for this business.

△ 9. A long watering trough has the cross section shown in the diagram below. Find the angle x that will maximize the capacity of this trough.

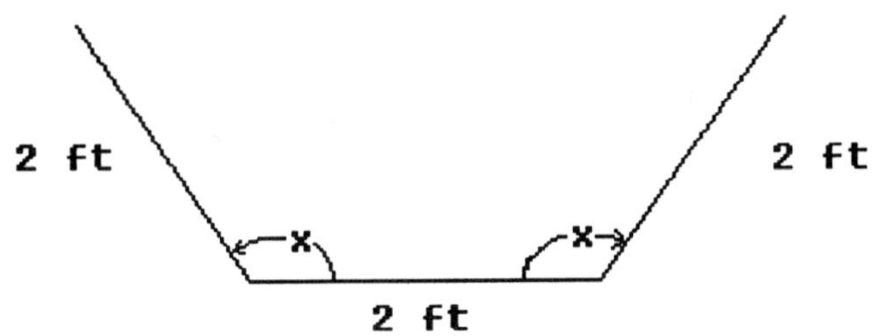

10. Maria applied for admission to a large French university. In order to be accepted she had to pass a test in French and German vocabulary. It so happened that a variable length five-to-ten week summer course in French was offered at Mill Creek Community College. MCCC advertised that for the average student with some previous exposure to French, the number of words w known after x weeks would be approximately

$$w(x) = 2100 - 2000e^{-0.3x}.$$

At the end of the course, most students would know about 2,100 words. Maria recognized this as a learning curve. She knew that she also had a forgetting curve for her German vocabulary. She had just finished a course in German, and she estimated that the number of German words u that she would remember after x weeks would be about

$$u(x) = 3000e^{-0.025x}.$$

Find the maximum number of words that Maria could expect to know and the optimal time for her to schedule the vocabulary examination.

⋆ 11. One day as Slick Rock was walking home from work, he noticed that a large tent had been erected in a vacant lot and that a revival meeting was being held. As he stopped to listen, he heard an amazing story. The speaker had been a missionary stationed on the north side of a large lake in a region populated by cannibals. The story involved the missionary's journey to the mission station. A guide took him to the south side of the lake, and showed him a small rowboat. He was to find his own way from there. The north side of the lake appeared to be a straight line running east and west. The mission station was 0.25 miles across the lake and 0.5 miles to the west. The missionary estimated that he could row at a rate of 1 mph, and he could walk through the thick jungle on the north side at a rate of 2 mph. At first he thought that he might take the boat directly to the mission station, but then he noticed the huge crocodiles in the water. He then determined to row directly to some point on the shore, and then walk the rest of the way through the cannibal infested jungle. Where should he land on the north side of the lake in order to minimize his travel time?

Chapter 8

Integrals

8.1 Limits of Sums

When Maria Garcia was in college, she was commissioned to paint the backdrop for the stage used in a rock concert. Her fee was to be based on the area that was to be painted. After some measurements were taken, she realized that the backdrop had the shape of the region bounded by the x-axis, the lines $x = 0$ and $x = 20$, and the graph of the function $f(x) = 6\log(x + 1)$ where the measurements were given in feet. She then determined to calculate the area of this backdrop.

At first she considered using a rectangle as an approximation. It seemed reasonable to use 20 yards as the width, but it was difficult to decide on a height. Finally, she chose $f(10) \approx 5.7255$ as the height. The estimate for the area was then $(20)(5.7255) = 114.51$ square feet. As she sat thinking about how to improve her estimate, Slick Rock walked up to where she was working. After she explained the problem, he said, "Why not use two rectangles?" This seemed like a splendid idea to Maria. She quickly drew the graph shown in Figure 8.1. She divided the interval [0,20] on the x-axis into two subintervals: [0,10] and [10,20]. These would form the bases for her two rectangles. She quickly (and somewhat arbitrarily) chose $f(5)$ and $f(15)$ for the heights of her two rectangles. Her new estimate for the area of the backdrop was

$$f(5)(10 - 0) + f(15)(20 - 10) \approx 3.6124(10) + 6.8768(10) = 104.8912.$$

Suddenly Maria had a brilliant idea. "We could get a better estimate by using even more rectangles," she said. "Consider using n rectangles. We could subdivide the interval $[0, 20]$ as follows: $0 = x_0 < x_1 < \cdots < x_n = 20$. Let's choose some number w_i in each subinterval $[x_{i-1}, x_i]$ to determine the height of ith rectangle. Then our approximation for the area will be

$$f(w_1)(x_1 - x_0) + f(w_2)(x_2 - x_1) + \cdots + f(w_n)(x_n - x_{n-1}).$$

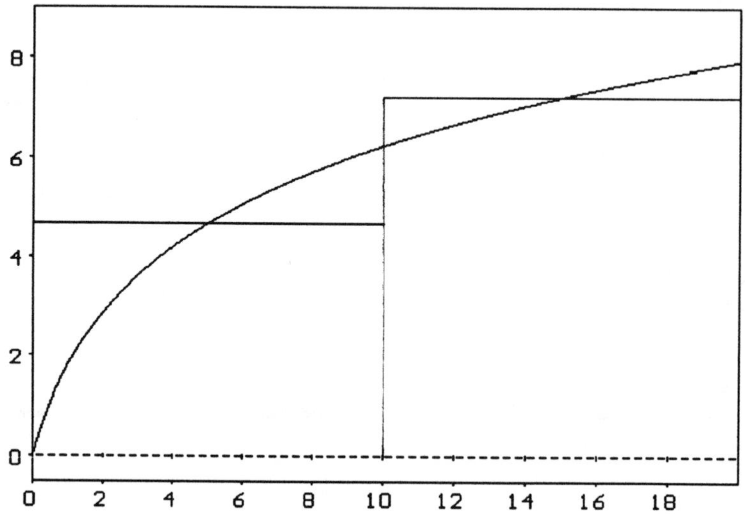

Figure 8.1: Approximating Area with Rectangles

Maria was an artist, and she decided that her formula would look better if she used the notation $\Delta x_i = x_i - x_{i-1}$. She remembered using the symbol Δx for *a change in* x in one of her science courses, so the use of the same notation in her formula seemed quite appropriate. Her new formula was

$$f(w_1)\Delta x_1 + f(w_2)\Delta x_2 + \cdots + f(w_n)\Delta x_n.$$

Slick Rock observed that the area of the backdrop must be the limit of this sum as n goes to infinity.

Maria decided to try her formula for 10 intervals of width 2 with w_i selected as the midpoint of the ith subinterval. Her calculations are summarized in the following table:

x_{i-1}	x_i	w_i	$f(w_i)$	$f(w_i)\Delta x_i$
0	2	1	1.80618	3.61236
2	4	3	3.61236	7.22472
4	6	5	4.66891	9.33782
6	8	7	5.41854	10.8371
8	10	9	6.00000	12.0000
10	12	11	6.47509	12.9502
12	14	13	6.87677	13.7535
14	16	15	7.22472	14.4494
16	18	17	7.53164	15.0633
18	20	19	7.80618	15.6124

The area is approximately 114.8, the sum of the numbers in the right column rounded to one decimal place. Sums of the form used by Maria are called *Riemann sums.*

8.1.1 Problem Solver Example

Use Problem Solver to improve upon Maria's estimate, by using 20, 40, 80, 160, and 320 subintervals. In each case, select w_i as the midpoint of the rectangle.

1. Start Problem Solver and select [Functions].
2. If you wish to print your results select [Options], [Output device], and finally [Printer]. Press the Esc key to return to the functions menu.
3. Select [Edit/enter functions] and select the first function.
4. Type the expression `6*log(x+1)` and press the Enter key.
5. Press the Esc key to return to the functions menu.
6. Select [Integrals] and then select the function that you entered.
7. Select [Lower limit of integration], and enter 0.
8. Select [Upper limit of integration], and enter 20.
9. Select [Algorithm] and then select [Riemann sums].
10. Select [Midpoint].
11. Select [Number of subintervals], and enter 20.
12. Select [Compute integral].
13. Change the number of subintervals appropriately and recompute the integral for 40, 80, 160, and 320 subintervals. The printed output follows:

```
Integrand:                                    6*log(x+1)
Limits of integration:                              0 20
Algorithm used:                 Riemann sums (midpoints)
Number of subintervals used:                          20
Value of the integral:                  114.582626888632
Estimated error bound:                  1.48552642589038
Estimated relative error:             0.0129646741938831

Integrand:                                    6*log(x+1)
Limits of integration:                              0 20
Algorithm used:                 Riemann sums (midpoints)
Number of subintervals used:                          40
Value of the integral:                  114.509775093603
Estimated error bound:                 0.464461550077033
Estimated relative error:            0.00405608647556394

Integrand:                                    6*log(x+1)
Limits of integration:                              0 20
Algorithm used:                 Riemann sums (midpoints)
Number of subintervals used:                          80
Value of the integral:                  114.490731753655
Estimated error bound:                 0.128505167419245
Estimated relative error:            0.00112240672629942

Integrand:                                    6*log(x+1)
Limits of integration:                              0 20
Algorithm used:                 Riemann sums (midpoints)
Number of subintervals used:                         160
Value of the integral:                  114.485907448424
Estimated error bound:                0.0333644423824949
Estimated relative error:           0.000291428378619661

Integrand:                                    6*log(x+1)
Limits of integration:                              0 20
Algorithm used:                 Riemann sums (midpoints)
Number of subintervals used:                         320
Value of the integral:                  114.484697131291
Estimated error bound:               0.00844237861761151
Estimated relative error:            7.37424199841292e-05
```

As the number of rectangles increases, it appears that the approximations to the area are approaching about 114.5 square feet. Notice that Maria's first estimate using one rectangle was actually closer than the estimate using 2 rectangles. Usually, increasing the number of rectangles improves the estimate. In this example, the single-rectangle estimate was not improved upon until 20 rectangles were used. Besides the number of rectangles, the accuracy depends upon the placement of w_i in the interval $[x_{i-1}, x_i]$.

Definition 14 *Let f be a function defined on the interval $[a, b]$. Let*

$$a = x_0 < x_1 < \cdots < x_n = b$$

be a partition of the interval, and let $w_i \in [x_{i-1}, x_i]$ for $i = 1, 2, \ldots, n$. The expression

$$f(w_1)\Delta x_1 + f(w_2)\Delta x_2 + \cdots + f(w_n)\Delta x_n,$$

which can also be written in the form

$$\sum_{i=1}^{n} f(w_i)\Delta x_i,$$

is called the Riemann sum *of f corresponding to the partition of $[a, b]$ and the numbers $w_i \in [x_{i-1}, x_i]$ for $i = 1, 2, \ldots, n$.*

Definition 15 *Let f be defined on $[a, b]$. The* definite integral *of f from $x = a$ to $x = b$, denoted by $\int_a^b f(x)dx$, is the limit of the values of the Riemann sums of f as the number of rectangles goes to infinity, and the width of the widest rectangle goes to zero.*

From the example given in this section, we see that

$$\int_0^{20} 6\log(x + 1)\, dx \approx 114.5.$$

If $f(x) \geq 0$ for $x \in [a, b]$, then $\int_a^b f(x)dx$ can be interpreted as the area under the graph of $y = f(x)$, above the x-axis, and between $x = a$ and $x = b$. However, even if $f(x)$ is negative for some values of x, it is still possible to calculate Riemann sums and the integral of f. In this case they represent something other than area.

Exercise Set 8.1

1. Evaluate the Riemann sum of $f(x) = x^3 - 12x + 2$ on the interval $[1, 4]$ using
   (a) 20 subintervals of equal length with w_i selected as the midpoint of $[x_{i-1}, x_i]$.
   (b) 20 subintervals of equal length with w_i selected as the left end-point of $[x_{i-1}, x_i]$.
   (c) 20 subintervals of equal length with w_i selected as the right end-point of $[x_{i-1}, x_i]$.

(d) 40 subintervals of equal length with w_i selected as the midpoint of $[x_{i-1}, x_i]$.

(e) 200 subintervals of equal length with w_i selected as the midpoint of $[x_{i-1}, x_i]$.

2. Approximate each of the following integrals using Riemann sums with w_i selected as the midpoint of $[x_{i-1}, x_i]$ and n rectangles.

   (a) $\int_0^1 \frac{x^2}{1+x^2} dx$, $n = 10$

   (b) $\int_{-2}^2 \frac{x^3}{1+x^2} dx$ $n = 20$

   (c) $\int_0^1 x dx$ $n = 25$

   (d) $\int_0^1 x^2 dx$ $n = 15$

   (e) $\int_0^1 x^3 dx$ $n = 22$

   (f) $\int_0^1 x^4 dx$ $n = 30$

   (g) $\int_0^1 x^5 dx$ $n = 100$

△ 3. Approximate each of the following integrals using Riemann sums with w_i selected as the midpoint of $[x_{i-1}, x_i]$ and n rectangles.

   (a) $\int_0^\pi \sin x dx$, $n = 10$

   (b) $\int_{-2}^2 \cos(x^2) dx$ $n = 20$

   (c) $\int_0^{\pi/4} \tan x dx$ $n = 25$

   (d) $\int_{-\pi}^\pi \sin^3 x dx$ $n = 15$

4. Use Riemann sums to estimate the area bounded by the x-axis, the lines with equations $x = 2$ and $x = 4$, and the graph of $f(x) = 2^x$.

5. Use Riemann sums to estimate the area of the top half of the circle with equation $x^2 + y^2 = 4$.

6. For each of the following linear functions, estimate some definite integral using Riemann sums. Experiment with the number of subintervals and the placement of w_i in the interval $[x_{i-1}, x_i]$. Draw some conclusions about where w_i should be placed for the greatest accuracy. What is the best accuracy that you can expect for linear functions? Justify your claims.

   (a) $f(x) = 3x + 4$

   (b) $f(x) = 7 - x$

   (c) $f(x) = 1 + 8x$

   (d) $f(x) = x$

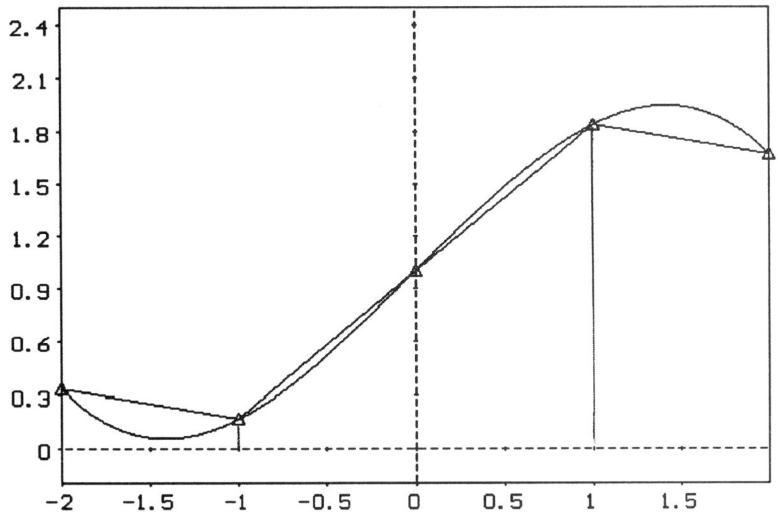

Figure 8.2: The Trapezoidal Rule

8.2 Simpson's Rule and the Trapezoidal Rule

Again consider finding an approximation for $\int_a^b f(x)\,dx$. For now we shall again assume that $f(x) \geq 0$ for $x \in [a,b]$ so that we are talking about the area of the region above the x-axis, below the graph of $y = f(x)$, and between $x = a$ and $x = b$. Rather than using rectangles to estimate the area, we will use trapezoids. Recall that the area of a trapezoid is the product of the height times the average of the bases. Thus the area of the trapezoid with vertices $(a,0)$ $(b,0)$, $(a, f(a))$, and $(b, f(b))$ is

$$\frac{b-a}{2}\left(f(a) + f(b)\right).$$

For more precision, we subdivide the interval $[a,b]$ into n subintervals of width $h = (b-a)/n$ using the partition $a = x_0 < x_1 < \cdots < x_n = b$ where $h = x_i - x_{i-1}$. The area of the region is now approximately the area of n trapezoids, namely

$$\frac{h}{2}(f(x_0) + f(x_1)) + \frac{h}{2}(f(x_1) + f(x_2)) + \cdots + \frac{h}{2}(f(x_{n-1} + f(x_n)).$$

This simplifies to the *trapezoidal rule*

$$\int_a bf(x)\,dx \approx \frac{h}{2}(f(x_0) + 2f(x_1) + 2f(x_2) + \cdots + 2f(x_{n-1}) + f(x_n)).$$

This formula is often more accurate than Riemann sums. A graphical representation of this method is given in Figure 8.2. The area under the curve is being approximated by the area under the broken line.

8.2.1 Problem Solver Example

We will use Problem Solver to estimate

$$\int_0^4 f(x)\, dx$$

with f defined by the following table:

x	0.00	0.50	1.00	1.50	2.00	2.50	3.00	3.50	4.00
$f(x)$	2.50	2.75	2.60	2.51	2.35	1.95	1.50	1.60	1.80

1. Start Problem Solver and select [Functions].
2. Select [Enter/edit functions] and select the second function.
3. This function will be entered as a table of ordered pairs. Select [Table of values].
4. There are 9 ordered pairs in this table. Enter 9 when this number is requested.
5. When a name for your table is requested, type `PS Example` and press the Enter key.
6. You should now be viewing the matrix editor. Enter the x-values in the first column and the $f(x)$ values in the second column. When you are finished, press the Esc key to leave the matrix editor.
7. Now press the Esc key again to return to the functions menu.
8. Select [Integrals] and then select the function that you entered.
9. Select [Compute integral]. The value should be 8.705.

Now consider the subintervals in groups of two and replace the graph of f with the parabola that passes through $(x_i, f(x_i))$, $(x_{i+1}, f(x_{i+1}))$, and $(f(x_{i+2}, f(x_{i+2}))$ as shown in Figure 8.3. This leads to Simpson's rule:

$$\int_a bf(x)\, dx \approx \frac{h}{3}(f(x_0)+4f(x_1)+2f(x_2)+4f(x_3)+3f(x_4)\cdots+2f(x_{n-2})+4f(x_n)+f(x_n)).$$

Simpson's rule is usually more accurate than either the trapezoidal rule or Riemann sums. Because the subintervals are used two at a time, this rule requires an even number of subintervals.

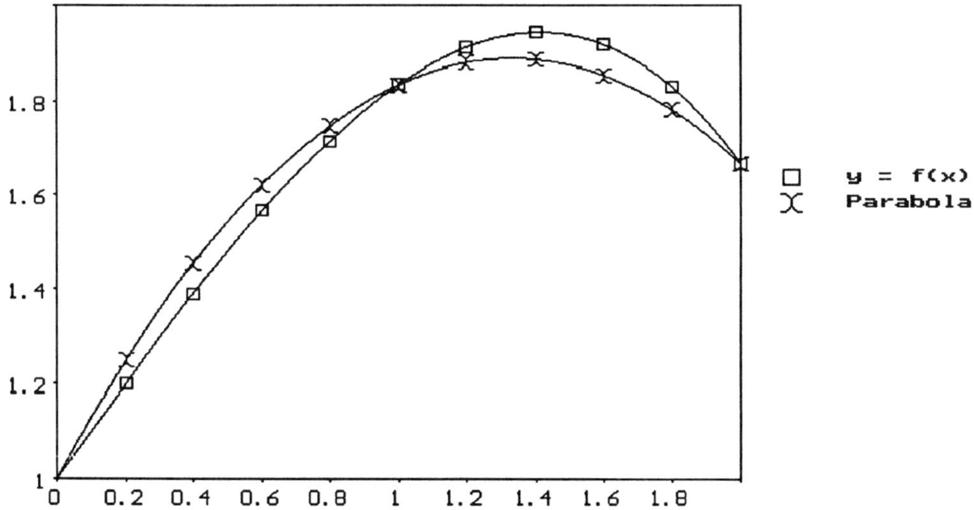

Figure 8.3: Simpson's Rule

8.2.2 Problem Solver Example

We will use Simpson's with rule with 50 subintervals to approximate

$$\int_{-2}^{3} \frac{1}{1+x^4}\, dx.$$

1. Start Problem Solver and select Functions.
2. Select Enter/edit functions and then select the first function.
3. Type the expression `1/(1+x^4)` and press the Enter key.
4. Press the Esc key to return to the functions menu.
5. Select Integrals and then select the function that you just entered.
6. Select Lower limit of integration and enter `-2`.
7. Select Upper limit of integration and enter `3`.
8. Select Algorithm and then select Simpson's rule.
9. Select Number of subintervals and enter `50`.
10. Select Compute integral. The value should be approximately 2.169.

Integrals can be found analytically using the following theorem, which is called the *Fundamental Theorem of Calculus.*

Theorem 9 *If f is continuous on an interval $[a, b]$, and if F is a function whose derivative is f, then*

$$\int_a^b f(x)\,dx = F(b) - F(a).$$

Unfortunately it is often difficult or impossible to find the needed function F.

Exercise Set 8.2

1. Approximate each integral using both Simpson's rule and the trapezoidal rule with n subintervals.

   (a) $\int_{-2}^{3} 2^x - 3\,dx$, $n = 10$

   (b) $\int_0^2 \log(x+1)\,dx$, $n = 20$

   (c) $\int_{-1}^{1} x^3 + x\,dx$, $n = 20$

   (d) $\int_0^1 x^2 + 1\,dx$, $n = 24$

△ 2. Approximate each integral using both Simpson's rule and the trapezoidal rule with n subintervals.

   (a) $\int_{-2}^{3} 2^{\sin x}\,dx$, $n = 10$

   (b) $\int_0^{\pi} \sin x\,dx$, $n = 20$

   (c) $\int_{-1}^{1} \sin^3 x + x\,dx$, $n = 20$

   (d) $\int_0^1 \tan(x^2) + \sec x\,dx$, $n = 24$

3. Use both the trapezoidal rule and Simpson's rule to estimate $\int_0^2 f(x)\,dx$ if f is defined by the following table:

x	0.00	0.25	0.50	0.75	1.00	1.25	1.50	1.75	2.00
$f(x)$	1.50	1.75	2.00	1.75	1.60	1.55	1.50	1.60	1.80

4. In a normal distribution with mean μ and standard deviation σ, the probability that a randomly selected data value will be between a and b is given by the integral

$$\int_a^b \frac{1}{\sqrt{2\pi}\sigma} e^{-\frac{1}{2}\left(\frac{x-\mu}{\sigma}\right)^2} dx$$

where $e \approx 2.71828\ldots$. Enter e^x into Problem Solver as `e^x` or as `exp(x)` . The latter form is more efficient. Use integrals to answer each of the following questions:

(a) The tread life of a certain type of tire has a normal distribution with a mean of 40,000 miles and standard deviation of 10,000 miles. Find the probability that a set of these tires will wear out between 30,000 and 45,000 miles.

(b) Standard IQ test scores have a mean of 100 with a standard deviation of 15. What percent of the population would be expected to have IQs between 85 and 115?

5. Use Problem Solver and Simpson's rule to estimate the area of the surface of the pond shown below:

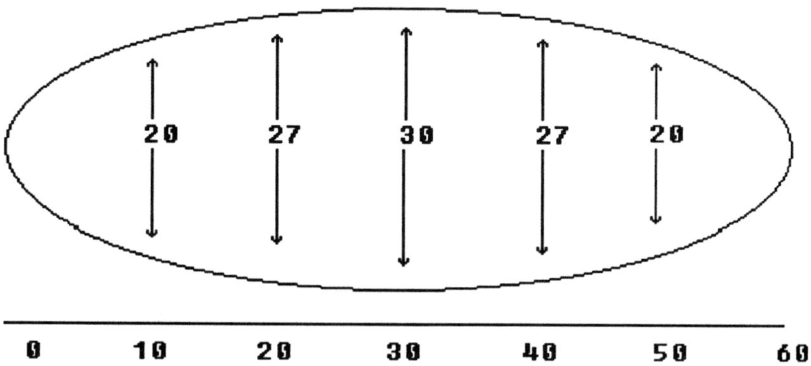

6. The length of the curve defined by the equation $y = x^3$ between the points $(0,0)$ and (a, a^3) is given by the integral

$$\int_0^a \sqrt{1+9x^4}\, dx.$$

(a) Find the length of the curve from $(0,0)$ to $(2,8)$.

(b) Find the length of the curve from $(0,0)$ to $3,27)$.

7. Use Problem Solver to apply the trapezoidal rule to several linear functions. Compare the results with those obtained by using the Fundamental Theorem of Calculus. What can be said about the error in the trapezoidal rule approximations? Explain the results.

8. Use Problem Solver to apply Simpson's rule to several quadratic functions (polynomials of degree 2). Compare the results with those obtained by using the Fundamental Theorem of Calculus. What can be said about the error in the Simpson's rule approximations? Explain the results. For a surprise, try

a collection of third degree polynomials. Then try a few fourth and higher degree polynomials.

★ 9. Approximate the area bounded by the graphs of $y = x + 1$ and $y = x^2$.

★ 10. Approximate the area bounded by the graphs of $y = x^4 + 1$ and $y = e^x$. Use Problem Solver first to find the intersection points and then to estimate the integral.

Appendix A

Answers to Odd-Numbered Exercises

Chapter 2

Section 2.2

1. $3, -3$

3. $1.302776, -2.302776$

5. 1.1486984

7. $(2{,}1), (-1.5, 2.75)$

9. $\pm 0.382683, \pm 0.923880$

11. -2.324718

Section 2.3

1. $-2.330059, 0.201640, 2.128419$

9. 143.75 seconds

Section 2.4

5. $f(x) = \begin{cases} 1.078x & \text{if } 1000 \leq x \leq 5000 \\ 1.088x & \text{if } 5000 < x < 10000 \\ 1.1x & \text{if } x \geq 10000 \end{cases}$

Chapter 3

Section 3.1

1. $x = 1$, $y = -1$, $z = 1$

3. The coefficient matrix is singular. There are multiple solutions: z is arbitrary, $y = 3 - 2z$, $x = -2 + z$. To get these solutions have the intermediate calculations displayed and then study the system represented by the last matrix.

5. $w = -0.135897$, $x = 0.584615$, $y = 4.68974$, $z = 2.87949$

7. $y = x^2 + 2x - 1$

9. \$132,183.91, \$74,712.64, \$93,103.45

11. \$78260.87, \$1,739.13, \$20,000

Section 3.2

1. (a) $5A = \begin{pmatrix} 10 & 15 & 35 \\ 25 & 5 & 40 \end{pmatrix}$

   (b) $A + B = \begin{pmatrix} 9 & 9 & 8 \\ 1 & 4 & 10 \end{pmatrix}$

   (c) $A - B = \begin{pmatrix} -5 & -3 & 6 \\ 9 & -2 & 6 \end{pmatrix}$

   (d) $3A - 7B = \begin{pmatrix} -43 & -33 & 14 \\ 43 & -18 & 10 \end{pmatrix}$

3. $$A^{-1} = \begin{pmatrix} 25 & -300 & 1050 & -1400 & 630 \\ -300 & 4800 & -18900 & 26880 & -12600 \\ 1050 & -18900 & 79380 & -117600 & 56700 \\ -1400 & 26880 & -117600 & 179200 & -88200 \\ 630 & -12600 & 56700 & -88200 & 44100 \end{pmatrix}$$

5. (a) FIND BATMAN (b) CONGRATULATIONS

Section 3.3

1. (a) $Z = 16.6667$ at $x_1 = 3.3333$ and $x_2 = 0$ (b) $Z = 15$ at $x_1 = 0$ and $x_2 = 2.5$ (c) There are no feasible solutions. (d) There is no solution. The objective function is unbounded.

3. Put \$33,333.33 in the 8% account, \$61,666.67 in the 11% account, and \$5000 into Ace stocks.

5. Ship 300 from factory 2 to the Oregon warehouse, 500 from factory 3 to the Oregon warehouse, 600 from factory 1 to the California warehouse, and 400 from factory 2 to the California warehouse.

7. Consider the fact that Slick's optimal diet selections are still available. Adding one more item simply gives him more options.

9. Build 3000 in the winter and install them all in the winter. Build 2800 in the spring. Install 1500 of them in the spring, 800 in the summer, and 500 in the fall. Build 3000 in the summer and install them in the summer (along with the 800 built in the spring for summer installation. Also take 700 out of storage and install them in the summer. Build 3000 in the fall for installation in the fall (along with the 500 built in the spring for fall installation). This leaves 500 in storage for next year. The total of the production and storage costs for this schedule is \$9,310,000.

Chapter 4

Section 4.1

1. (a) 40320 (b) 3024 (c) 15504 (d) 1260

3. $2.6313 \cdot 10^{35}$

5. 12,279,168

Section 4.2

1. $4.91 \cdot 10^{-4}$

3. $1.705 \cdot 10^{-15}$

5. (a) $1.847 \cdot 10^{-5}$ (b) 0.0253 (c) 0.00858 (d) $2.40 \cdot 10^{-4}$ (e) 0.00144 (f) 0.00198 (g) 0.00512 (h) $2.00 \cdot 10^{-5}$ (i) $1.539 \cdot 10^{-6}$

Section 4.3

1 (a) 0.176197 (b) 0.736824 (c) $9.5367 \cdot 10^{-7}$ (d) 0.999999

3. 0.317652

5. 0.988528

Section 4.4

1 (a) 0.135905 (b) 0.682689 (c) 0.5 (d) 0.308538 (e) 0.00620967

3. 0.237525

5. 0.0912112

Section 4.5

1 (a) 8,8,0 (b) 8,8,0.57735 (c) 8,8,11.3137

Section 4.6

1 (a) $y = 0.5x + 1.5$ (b) $y = 0.75x + 1.08333$ (c) $y = 0.7x + 1.2$

3. $y = 0.00194179x + 3.27524$

Section 4.7 Since the data for these problems are randomly generated, the answers will vary.

Chapter 5

Section 5.1

1. $3, 5, 7, 9 \ldots, 101$

3. \$531,441

5. \$124,845.83, \$124,690.25, \$124,533.28,...,\$120,882.43

Section 5.2

1. (a)(i) 1.464 (a)(ii) 1.596 (a)(iii) 1.635 (a)(iv) 1.640 (b) They seem to be converging to 1.640.

3. \$7.63, \$7.66, \$7.70,...,\$8.57 for a total of \$194.11.

Section 5.3

3. (a) \$2,160.00, \$2,332.80, \$2,519.42,...,\$7,400.04 (b) 7.27 years

5. (a) 0.94 days (b) 96.02

7. (a) 100 (b) 10,000 (c) 10,000,000 (d) 100,000,000 (e) 316,227,766

Section 5.4

5. 2.696257, 7.328348, 8.263830

Chapter 6

Section 6.1

1. (a)

x	$f(x)$
2	3
4	5
8	9
⋮	
1024	1025

(b)

x	$f(x)$
10	11
20	21
30	31
⋮	
120	121

(c)

x	$f(x)$
−5	−4
−10	−9
−15	−14
⋮	
−100	−99

(d)

x	$f(x)$
1	2
−1	0
1	2
⋮	
1	2

3. (a)

x	$f(x)$
1	1
2	4
3	7
⋮	
15	43

(b)

x	$f(x)$
10	28
20	58
30	88
⋮	
100	298

(c)

x	$f(x)$
5	13
10	28
15	43
⋮	
125	373

(d)

x	$f(x)$
0	−2
−1	−5
−2	−8
⋮	
−12	−38

5. (a) −1 (b) doesn't exist (c) −2 (d) 1

7. (a) 0.75 (b) 0.167 (c) 0 (d) 0

9. (a) 1 (b) 0 (c) 0 (d) doesn't exist

11. $142.63

Section 6.2

1. (a) 6 (b) 0.245 (c) 0.0412 (d) 0.0048

3. (a) 7.389 (b) 0.028824 (c) 0.0048445 (d) 0.00569945

5. (a) 0 (b) 0.5 (c) 0.1715 (d) 0.0201768

7. (a) 0 (b) 0.000814207

9. (a) 1 (b) 0.5 (c) 0.5 (d) 0.245

11. (a) continuous (b) not continuous (c) not continuous (d) continuous

Chapter 7

Section 7.1

1. (a) 0.250 (b) 0.275 (c) 0.333 (d) 0.375 (e) 0.500

3. (a) 0 (b) 4 (c) 13 (d) 0.289 (e) -0.111 (f) 0.145

5. (a)

x	$f(x)$	$f'(x)$	$f''(x)$
1.0	5.0	0.0	0.0
1.2	5.0	0.0	0.0
1.4	5.0	0.0	0.0
⋮			
5.0	5.0	0.0	0.0

(b)

x	$f(x)$	$f'(x)$	$f''(x)$
1.0	1.0	3.0	0.0
1.2	1.6	3.0	0.0
1.4	2.2	3.0	0.0
⋮			
5.0	13.0	3.0	0.0

(c)

x	$f(x)$	$f'(x)$	$f''(x)$
1.0	−1.0	−4.0	0.0
1.2	−1.8	−4.0	0.0
1.4	−2.6	−4.0	0.0
⋮			
5.0	−17	−4.0	0.0

(d)

x	$f(x)$	$f'(x)$	$f''(x)$
1.0	2.00	4.0	4.0
1.2	2.88	4.8	4.0
1.4	3.92	5.6	4.0
⋮			
5.0	50.00	20.0	4.0

(e)

x	$f(x)$	$f'(x)$	$f''(x)$
1.0	2.718	2.676	2.594
1.2	3.320	3.342	3.331
1.4	4.055	4.082	4.069
⋮			
5.0	148.413	146.705	144.016

(f)

x	$f(x)$	$f'(x)$	$f''(x)$
1.0	0.000	0.426	−0.388
1.2	0.079	0.365	−0.306
1.4	0.146	0.312	−0.224
⋮			
5.0	0.699	0.087	−0.017

(g)

x	$f(x)$	$f'(x)$	$f''(x)$
1.0	1.000	−0.952	1.637
1.2	0.833	−0.714	1.190
1.4	0.714	−0.521	0.744
⋮			
5.0	0.200	−0.040	0.016

Section 7.2

1. (a) 3 (b) 2 (c) 0.735 (d) 6

3. (a) 3 (b) 3.184387 (c) 1, −1 (d) −0.618034, 0, 1.618034 (e) −0.866025,0,0.866025

5. (a) 1.934563 (b) 1.258728 (c) 0.739085 (d) ±0.890553

Section 7.3

1. (a) Increasing on $[1.5, \infty)$ and decreasing on $(-\infty, 1.5]$. (b) Increasing on $[1, \infty)$ and decreasing on $(-\infty, 1]$. (c) Increasing on $(-\infty, -2]$ and $[2, \infty)$. Decreasing on $[-2, 2]$. (d) Increasing on $[0, \infty)$ and decreasing on $(-\infty, 0]$.

3. (a) Increasing on [3,5] and decreasing on $[-5, 3]$. (b) Increasing on $[-1, 0]$ and [0.7,3]. Decreasing on [0,0.6]. The table gives no information about [0.6,0.7] except that somewhere in the interval the function changes from decreasing to increasing. (c) Increasing on $[-0.8, 1.6]$. Decreasing on $[-10, -0.9]$ and [1.7,10]. (d) Increasing on [0.9,7.3]. Decreasing on [0,0.8] and [7.4,10].

5. (a) minimum $f(-1) = -1$, maximum $f(2) = 26$ (b) minimum $f(0) = -1$, maximum $f(-2) = 7$ (c) minimum $f(2) = -244$, maximum $f(0.931349) = 5.095350$ (d) minimum $f(5) = -3$, maximum $f(-5) = 28.96875$

7. 22.25 by 22.25 ft.

9. 2.094395 radians

11. He should come ashore 0.3557 miles east of the mission station.

Chapter 8

Section 8.1

1. (a) -20.2922 (b) -22.1906 (c) -18.1406 (d) -20.2605 (e) -20.2504

3. (a) 2.0082 (b) 0.912706 (c) 0.346532 (d) $2.099 \cdot 10^{-16}$

5. 6.2834 with 499 subintervals

Section 8.2

1. (a) -3.818230, -3.707422 (b) 0.562774, 0.562534
   (c) $1.776 \cdot 10^{-16}$, $2.220 \cdot 10^{-16}$ (d) 1.333333, 1.333623

3. 3.35, 3.341667

5. 1293.333

9. 1.863390

Appendix B

Personal Computers and DOS

B.1 The Hardware

Most personal computer (PC) systems consist of three main parts: the system unit, keyboard, and video monitor.

The system unit contains:

- The central processing unit (CPU), which is ordinarily contained on a single chip such as the Intel 8088 used in the original IBM-PC, along with its retinue of support chips. If a computer system has a brain, it is located in the CPU. It is the single most important piece of electronics in a computer.

- Read-only memory (ROM) containing permanent data and programs that are used by the programs you will run.

- Random-access memory (RAM). The programs that you will run on your computer are transferred from a disk to this type of memory. After the programs have finished running, the RAM is made available for other programs that you might wish to run.

- Disk drives. Your programs and data are stored on some type of disks. The 3.5 inch and 5.25 inch flexible plastic disks will be referred to as *floppy disks*. The disks permanently installed in many computers are made of metal and are called *hard disks*. Many of these have a storage capacity of 50 or more floppy disks. If you have a single floppy disk drive, it is designated as drive A. If you have two, the top or left one is drive A, and the bottom or right one is drive B. If you have a hard disk, it is probably called drive C.

B.2 The Operating System

DOS is an acronym for *disk operating system.* The operating system acts as a supervisor for your computer system by handling the loading and execution of programs and coordinating communication between the system unit and the video screen, disk drives, keyboard, and other peripheral devices.

In addition to DOS, your computer has some programs built into the ROM. When you turn on your computer, it knows just enough to load DOS from a disk. Only after it has loaded DOS is it able to run programs. Computers use a set of instructions called a *bootstrap loader* to load an operating system. It is as if the computer were pulling itself up by its own bootstraps. This colorful terminology is carried into such expressions as "booting a computer" or "formatting a boot disk."

Using computers can at times be frustrating, and the expression "booting a computer" can evoke dramatic if incorrect visual images of users taking revenge on these cantankerous machines! When you start your computer, you are booting the computer.

B.2.1 Starting Your Computer

Systems with a hard disk can be started by simply turning on the power switch. Systems without a hard disk are only slightly more complicated. A DOS disk was supplied with your operating system. Place this disk in drive A. If the drive has a door or lock, close it to secure the disk in place. Then turn on the power switch.

B.2.2 Restarting Your Computer

When a program goes so awry that the computer seems to be in gridlock, you can usually restart or reboot the computer by pressing the Ctrl, Alt, and Del keys simultaneously. This is a called a *warm boot.* If that will not restart the computer, your last resort is to turn off the computer, wait about a minute, and then turn it back on again. This process is called a *cold boot.* Warm boots are preferable to cold boots since cold boots can subject electronic components to stress. Some computers have reset switches. If your computer has one, use it. This is better than turning the computer off and then back on.

B.2.3 DOS Commands

DOS includes a command processor. It interprets and executes commands entered from the keyboard.

Changing the Current Drive

The current drive is the disk drive that is active at the moment. When your computer is turned on, but you are not running any programs, the letter designation for the current drive appears on the left side of the monitor in what is called the DOS prompt. If the DOS prompt is `C>`, the current drive is drive C. To change the current drive to drive A, type

```
A:
```

and then press the Enter key.

Formatting Floppy Disks

All floppy disks used in personal computers must be formatted. The formatting process places magnetic signals on the disk that help DOS store information in an orderly manner. This makes it possible for DOS to find the information when it is needed. You might think of the formatting process as placing a road map on the floppy disk. This map tells DOS where information is stored on the disk.

To format a data floppy disk in drive A, use the following steps:

1. Place your DOS floppy disk in drive A. This disk was included with the operating system for your computer.

2. Type `FORMAT A:` and press the Enter key.

3. Remove the DOS disk and replace it with a blank disk. Now follow the instructions given on the monitor.

You will see the disk drive light come on for a few minutes while DOS works at the formatting job. CAUTION: If you have a hard disk in your system, the importance of including "A:" in the above command cannot be over emphasized. If you omit "A:", you could accidentally reformat your hard disk and lose all of the information stored on it.

At times you may wish to format a *system disk* or *boot disk*. DOS files are placed on these disks making it possible for you to start your computer with them in place of the DOS disk. For example, if you create a boot disk for Problem Solver, you will be able to both start your computer and run Problem Solver using only your Problem Solver floppy disk. The command for formatting a system floppy disk is

```
FORMAT A:/S
```

Getting a Directory Listing

Programs and data are stored on floppy disks in files. The Problem Solver program consists of the file PS.EXE and numerous help files. To see the list of the files on your Problem Solver floppy disk, place the Problem Solver in drive A, change the current drive to drive A, type DIR, and press the Enter key.

File Names

DOS file names consist of up to 8 characters. If you like, the filename can be followed by a period and an extension of up to 3 characters. The extensions are useful for categorizing your files. For example, you might use .DAT as an extension on all of your data files for PS. All of the help files on your PS help disk have .HLP as an extension. The characters making up the filename can be letters or numbers as well as some other characters. See your DOS manual if you wish to use characters other than letters and numbers. Here are a few examples of file names:

MATRIX.DAT
FUNCTION.DAT
AUTOEXEC.BAT
PS.EXE
12345678.abc
MYFILE

Spaces are never allowed in file names.

Copying Files

Occasionally you will need to copy files from one location to another. You may wish to transfer files from one floppy disk to another or from a floppy disk to a hard disk. Use the COPY command to accomplish this. For example, if you want to copy the file MATRIX.DAT from drive A to drive B, type the command

```
COPY A:MATRIX.DAT B:
```

and press the Enter key. If you want the copy to have the name MATRIX2.DAT, type the command

```
COPY A:MATRIX.DAT B:MATRIX2.DAT
```

and press the Enter key.

Deleting Files

A normal part of computer housekeeping is deleting files that are no longer useful. Use the DEL command to do this. Examples follow:

```
DEL MATRIX.DAT
DEL DAT123.456
```

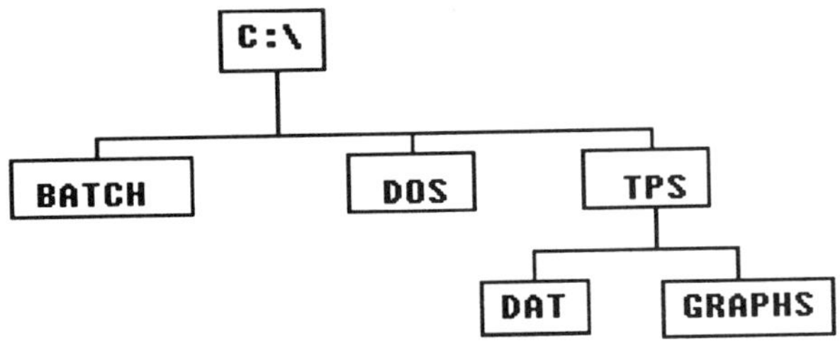

Figure B.1: Tree-Structured Directory

Using Subdirectories

Subdirectories are marvelous for organizing the files on a hard disk. Imagine trying to find one file among several thousand files on a hard disk. If we separate the files into groups, it will be easier to find the files when we need them. Subdirectories are groups of files that are ordinarily related in some way. For example, you might want to put the Problem Solver program and all of its help files into a subdirectory. This will keep the Problem Solver files separate from the other files on your hard disk. Placing a group of related files in their own subdirectory is not unlike taking a group of related file folders (perhaps those dealing with correspondence) and placing them in their own file drawer.

Make a subdirectory with the MDIR command. To make a directory called TPS, type

```
MDIR TPS
```

and press the Enter key.

Subdirectories can be created within subdirectories to form a tree structure of subdirectories such as the one shown in Figure B.1. The highest level directory is called the *root directory*. It is denoted by `C:\`.

File Specifications

A file specification consists of a file name with a prefix, which gives the location of the file in the directory structure. This prefix begins with a drive letter followed by a colon, and it ends with a sequence of subdirectory names separated by the backward slash character. For example, suppose that the file GRAPH.DAT resides in the DAT subdirectory, which is defined within a subdirectory called TPS in drive C as shown in Figure B.1. The file specification for this file is

```
C:\TPS\DAT\GRAPH.DAT
```

Changing the Current Subdirectory

The current subdirectory is the subdirectory that is active at the moment. Typing `DIR` gives a list of the files in the current subdirectory. The CDIR command is used to change the current subdirectory. If the current subdirectory in the above example is TPS, you can switch to the subdirectory DAT by typing

```
CDIR C:\TPS\DAT
```

and pressing the Enter key. To change to the root directory, type

```
CDIR \
```

and press the Enter key.

Removing Subdirectories

At times you will want to remove subdirectories. Use the RDIR command to do this. Consider removing the subdirectory DAT shown in Figure B.1. First change the current directory to the subdirectory one level above DAT using the command

```
CDIR C:\TPS
```

and then use the command

```
RDIR DAT
```

Before a subdirectory can be removed, all of the files within the subdirectory must be deleted. In addition, all of the subdirectories within the one you wish to remove must themselves be removed. This means that a subdirectory must contain no files or subdirectories if it is to be removed.

Running Programs

To run a program stored on a floppy disk, use the following steps:

1. Place the floppy disk containing the program in drive A.
2. Change the current drive to drive A.
3. Type the name of the program and press the Enter key. If the Problem Solver Program disk is in drive A, and if you want to run Problem Solver, just type PS in response to the `A>` prompt, and press the Enter key.

To run a program stored on a hard disk, use the following steps:

1. Change the current subdirectory to the subdirectory containing the program that you wish to run.
2. Type the name of the program and press the Enter key.

Appendix C

Reference Guide for The Problem Solver

C.1 Precision and Limitations in Calculations

All calculations in Problem Solver are carried out with a precision of 15 to 16 significant digits. The upper limit for numbers is 10^{100}. Calculations leading to larger numbers cause an error message to be displayed.

If your computer has an 8087 or compatible numeric data coprocessor, Problem Solver will use it and run much faster. If your computer does not have a numeric data coprocessor, one will be emulated in software. This means that having or not having a coprocessor does not affect the accuracy of the calculated results. It affects only the speed at which Problem Solver runs.

C.2 The Help System

To get help while running Problem Solver, press the F1 key. Information pertinent to your current activity will be displayed on the screen. If you have a question on a certain topic, read the help screens available from the menus or edit screens related to that topic. Often the help available requires more than one screen for its display. If additional help is available, a PgDn indicator will be visible near the bottom of the screen. Use the PgDn and PgUp keys to move through the help screens.

To illustrate the use of the help system with a difficult problem, suppose that you have a question about how to enter a linear programming model into Problem Solver. Press the F1 key while only the main menu is on the screen. You will learn that linear programming problems are handled under the matrices menu. Select [Matrices], and when the matrices menu is on the screen, press the F1 key. Since the help screens available from the matrices menu do not answer this particular question, press the Esc key to leave the help system, and then select Linear programming, and again press the F1 key. The needed help, along with an example, is contained in these screens. The information on how to enter a linear programming model into Problem Solver is also available from the model editor. You can view this help by pressing the F1 key while you are viewing the editor screen.

Similarly, you can learn how to enter a system of equations into Problem Solver by selecting Linear equations/inverses, and then pressing the F1 key. Each menu with an "F1-Help" title has an associated help file. If you press the F1 key while viewing the help screens, the name of the help file being used will be displayed. You can add information to these files yourself. They are text files with a special format. Each line has at most 65 characters, and the first line and every twenty-first line thereafter are used for the titles, which appear at the top border of the help display. Thus 20 lines are available for each help display.

C.3 Command Line Parameters

Problem Solver can be started with any of the following command line parameters: ATT, BW, CGAHI, CGALO, EGAHI, V, and VGA. These parameters are separated by spaces and placed on the command line used to start Problem Solver. For example, Problem Solver could be started with the command

```
PS V CGAHI
```

The meanings of these parameters are given below:

ATT Use this parameter if you are using an AT&T 6300 computer. This activates the high resolution 640×400 graphics screen. If you do not use this command line parameter, your computer will display graphics in the lower resolution 640×200 CGA mode.

BW This option removes color from the display. For monochrome systems that display colors as varying shades of some single color, it is impossible to distinguish between some of the colors. Using this parameter makes the screens much more readable.

CGAHI Use this if you want your computer to display graphs in the CGA 640×200 monochrome mode. The most likely reason for using this parameter is to cure printing problems.

CGALO For computers equipped with the CGA card, the graphs will be displayed in the highest resolution available. This is the 640×200 monochrome mode. If you prefer the 320×200 color mode, use the command line parameter CGALO.

EGAHI If your system has an EGA graphics system with the minimum 64KB of graphics memory, Problem Solver will use the 16-color 640×200 mode. If you prefer the 4-color 640×350 mode, use the parameter EGAHI on the command line.

V Problem Solver directly modifies video memory in order to achieve increased speed in the video displays. If you want the operating system to handle the video output, use the parameter V on the command line. This will decrease the speed of the video output, but it may be helpful if you wish to run Problem Solver under windowing environments.

VGA Problem Solver should be able to detect a VGA system. However, if you have a VGA system, and if Problem Solver uses CGA resolution, use the parameter VGA to put Problem Solver in the VGA mode. In order to have the highest VGA resolution, you need to be using a version of DOS at least as new as version 4.0.

Several of the command line parameters modify the resolution of the graphics screens. This affects only the graphs; the text displays are always in the highest resolution available on your system.

C.4 Graphics Screens and Printed Graphs

Except in the case of AT&T computers, Problem Solver is able to detect and use the best graphics mode for your computer. This includes computers equipped with Hercules, CGA, EGA, and VGA systems.

There is probably only one situation where you would want to use the command line parameter CGAHI discussed in the last section. Problem Solver uses the program GRAPHICS.COM to help it print graphs. This program is part of the DOS that came with your computer. Versions of DOS older than 4.0 support the CGA modes and sometimes EGA modes, but only 4.0 or newer versions support the VGA mode. This means that if you are using a version of DOS older than 4.0, and if you have a VGA monitor, you can not use the highest possible screen resolution for your graphs. Problem Solver looks to see what version of DOS is being used, and then picks the highest screen resolution that will allow trouble-free printing. If in spite of this, you still have persistent printing problems, try starting Problem Solver with CGAHI or CGALO as a command line parameter as discussed in Section C.3.

The DOS program GRAPHICS.COM does not support Hercules-type graphics on monochrome monitors. If your computer is equipped with one of these graphics systems, graphs will be displayed on the screen, but you may not be able to print them. In order to print graphs, you would need to replace the DOS program GRAPHICS.COM with a similar program that supports your graphics monitor.

Operator Symbol	Operation
+	addition
-	subtraction
*	multiplication
/	division
^	exponentiation

Table C.1: Operations Available in Problem Solver

C.5 Files

Most of the Problem Solver menus have options for loading and saving files. To load a file, select the menu that is appropriate for your problem and select this option. For example, if you want to load the augmented matrix for a system of equations, first select [Matrices], then [Linear equations/inverses], and finally [Load].

After you select [Load], a window will open with a default directory name including a directory mask. You may do any of the following:

- Press the Enter key to select your file from a list of files.
- Type the name of your file and press the Enter key.
- Edit the given directory or the directory mask. For example, if the current directory is `C:\TPS\*.*`, and if you wish to keep this directory but view only the files with a `.DAT` extension, you should change the directory name to `C:\TPS\*.DAT`. When you press the Enter key, all of the files in this directory with a `.DAT` extension will be displayed, and you will be able to select any one of them.

You can even change the default disk drive by editing the directory name. If you replace the current directory with `A:`, the new directory will be the directory on drive A that was most recently active. Using `A:\` makes the root directory on drive A your default directory. After changing the default directory, all file operations are carried out in the new directory.

C.6 Mathematical Expressions

You can enter your own functions and expressions into Problem Solver using the operators shown in Table C.1. These operations are carried out in a specific order:

1. First, exponential expressions are evaluated from right to left.
2. Next, multiplication and division are done from left to right.

3. Finally, addition and subtraction are done from left to right.

In addition to the above operations, parentheses can be used for grouping. Expressions in parentheses are evaluated first. Examples follow:

```
   30 + 4*2  =  38  (multiplication first)
   (30+4)*2  =  68  (parentheses first)
     24/2*2  =  24  (left to right evaluation)
   24/(2*2)  =   6  (parentheses first)
      2^4/2  =   8  (exponents before division)
    2^(4/2)  =   4  (parentheses first)
2^3^2 = 2^9  = 512  (right to left for exponents)
```

A variety of scientific functions is available in Problem Solver. These include:

1. Trigonometric functions: $sin(x)$, $cos(x)$, $tan(x)$, $cot(x)$, $sec(x)$, and $csc(x)$.

2. Inverse trigonometric functions: $asin(x)$, $acos(x)$, and $atan(x)$.

3. Hyperbolic functions: $sinh(x)$, $cosh(x)$, and $tanh(x)$.

4. Logarithmic and exponential functions: $log(x)$, $ln(x)$, and $exp(x)$.

5. Miscellaneous functions:

$abs(x)$	–	absolute value of x
$sqr(x)$	–	square of x
$sqrt(x)$	–	square root of x
$step(x)$	–	0 if $x < 0$, and 1 otherwise
$gint(x)$	–	greatest integer less than or equal to x
$fact(n)$	–	n factorial
$p(n,r)$	–	number of permutations of n objects selected r at a time
$c(n,r)$	–	number of combinations of n objects selected r at a time
$rnd(x)$	–	random positive number less than 1. If the argument x is a positive integer, it is used as a seed to initialize the random number generator. Otherwise, the system timer is used to generate a seed.

The arguments of all functions must be enclosed in parentheses.

In addition to the above functions, Problem Solver recognizes two constants. You can use `pi` and `e` for the constants π and e. Problem Solver uses the values 3.141592653589793 and 2.718281828459045, respectively.

The following examples show expressions that can be entered into Problem Solver:

```
sin(2*x-1) + 3
2 + cos(sin(pi*x))
sqrt(5.8)-ln(x^2+1)
```

The default variable name is x, but it can be changed using the functions menu.

Expressions can be entered whenever numerical data is requested. For example, you could enter 1/3 rather than 0.3333333333333333. When you are entering numerical data, any of the scientific functions can be used, but no variables are allowed. When integers are expected for input, noninteger input is rounded to the nearest integer. Thus when asked how many rows your matrix is to have, you could enter something like 5.2. Problem Solver would round this to 5 and set up a matrix with 5 rows.

C.7 Setting Program Options

You can control many aspects of how Problem Solver runs by selecting [Options] from the main menu.

```
F1-help

Intermediate calculations displayed:  yes
Mode:  radians
[Format for text output]
Output device:  video screen
Calculator
Zero tolerance:  1e-12
Warning tone:  yes
User control over page breaks:  no
Save options
```

Using the options menu shown above, you can indicate a variety of preferences such as whether or not output is sent to a printer or whether a warning tone will sound when an incorrect key is pressed.

Intermediate Calculations Displayed: Selecting this item allows you to choose whether or not Problem Solver will display the intermediate steps involved in solving equations, estimating integrals, and solving linear programming problems. No intermediate steps can be displayed in the statistics routines.

Mode: Use this selection for changing back and forth between the degree and radian modes for trigonometric functions.

Format for Text Output: A number of options are available to help you produce attractive printed reports.

```
F1-help

Numerical field width in matrices:  12
Top margin (lines):  3
Lines to print on each page:  55
Left margin (number or spaces):  5
Characters to print per line:  65
Initialization codes for printer:
Header:
Line number for header:  2
```

Numerical Field Width in Matrices: If the entries in your matrices appear too crowded, choose a larger number here. If you want to see more entries on the screen at one time, select a smaller number. The field width should be large enough to accommodate both the numbers and any space that you wish to display between the numbers. If a number can not be displayed in the field width specified, Problem Solver will convert it to scientific notation and try again. If Problem Solver is still unable to display the number in the allotted space, it will display asterisks where the number should appear.

Decimal Places in Matrices: Select a larger number here if you want to see more decimal places in your matrices. Note that matrices are used for displaying solutions to systems of equations and linear programming problems as well as for a variety of data input. If you select fewer decimal places, your data is not actually rounded off. Problem Solver always works with 15 or 16 significant digits. Only the display is modified. If you increase the number of decimal places, you will probably also want to increase the field width for the numbers by an equal amount.

If you wish to display the number 12.654 with two spaces on the left, use a field width of 8 and enter 3 as the number of decimal places.

Top Margin: Enter the number of lines that should be skipped at the top of each page. If you use a header, it will appear on one of these lines.

Lines to Print on Each Page: If a full page is 66 lines, and if you have selected a 4 line top margin and want a 5 line bottom margin, set the number of lines per page to $66 - 4 - 5 = 57$.

Left Margin: Enter the number of blank spaces that you want on the left side of each line.

Characters per Line to Print: If your printer prints 80 characters per line and you want a 5 space left margin and a 4 space right margin, set the number of characters per line equal to $80-5-4 = 71$. If you can put your computer into a compressed print mode, you can print more characters per line.

Initialization Codes for Printer: See your printer manual for the decimal codes for any features that you wish to use. You might want to use the codes for compressed print for printing matrices. Enter your code numbers on a single line separated by spaces. For the IBM Graphics Printer, the code for placing the printer into compressed print is 15. To put this printer into compressed print, enter 15 on this line. These codes are sometimes given in three forms: ASCII, decimal, and hexadecimal. Use the decimal form of the codes.

The default codes set the line spacing to six lines per inch for IBM Grapics and Epson printers. When graphs are printed, the line spacing sometimes changes. These numbers are used to restore the spacing. Because of the obvious importance of maintaining the proper line spacing, you may wish to keep these codes and add additional codes at the end of the list of numbers.

Whenever a new problem is printed or a new page is started, these codes are sent to your printer.

Header: This line will appear at the top of each page of your text output but not on any graphs. If you use the symbols ## anywhere in your header, Problem Solver will replace these symbols by the current page number. For example, if you entered "Physics Laboratory page -##-," Problem solver would print this at the top of each page with the correct page number.

Line Number for Header: This line should be one of those ordinarily left blank for the top margin. For example, if you want 3 blank lines at the top of each page followed by your header and another blank line, set your top margin equal to 5. Then place the header in line 4.

We now return to discussing the choices in the options menu.

Output Device: Problem Solver displays all of its output on the screen. In addition, you can request that the output be sent to a printer or a file. If you send your output to a file, you will be able to load it into a word processor and edit it before printing. Only text output can be saved in this way. Graphs can not be saved in files.

Selecting the output device has no affect on calculations already performed. Results calculated *after* this selection will be sent to the printer or a file.

Calculator: The calculator option allows you to enter a mathematical expression. Problem Solver then calculates the value of your expression. Variables are not allowed in these expressions, but you may use any of Problem Solver's scientific functions.

Zero Tolerance: We are used to viewing numbers as either zero or nonzero. In the world of computers and calculators, things are not quite so simple. A problem arises when computers or calculators introduce small errors into calculations. Just as you can not represent 1/3 exactly as a decimal number in finitely many decimal places, a computer can not represent 0.1 exactly as a binary number in a finite amount of space. There are many other numbers that suffer the same fate when entered into a computer. The result is called round-off error, and it is unavoidable. These numbers with their small errors are then used in calculations, and the errors are carried into the results. We have all seen a calculator display of 3.9999999999999 when it should have been 4.

This makes it difficult to tell whether a number is actually zero or not. For example, is 10^{-20} the computer's approximation for zero, or should it be treated as a positive number?

The zero tolerance is used in the matrix calculations. Problem Solver will not divide by any number whose absolute value is less than this tolerance. Setting this number too small may cause Problem Solver to incorrectly report solutions for some problems that have no solution. On the other hand, setting this value too large may cause Problem Solver to incorrectly state that some problems have no solution when in fact they do.

Different problems may require a different value for this parameter. You could experiment by solving the same problem with various values of the zero tolerance.

Warning Tone: Problem Solver sounds a tone if your press an inappropriate key for menu responses and other user input. Select this option if you prefer to work in silence.

User Control over Page Breaks: Select this option for more printer control. If you answer **yes** to this question, you will be asked whether to start a new page at the beginning and end of each printed problem. Do not use the form feed button on the printer to advance the paper until you are completely finished printing. This causes Problem Solver to lose track of the number of lines printed on the current page. Consequently, Problem Solver will not know where to begin and end pages.

Save Selections: You may decide that you like your selections better than the default selections. By saving your selections, you will cause Problem Solver to automatically load and use your selections as its default selections the next time it runs. This configuration file will be saved in the current default directory.

Should you ever wish to return to the original default values, delete the file PSOPTION.CTL. Do this from the DOS command line rather from within Problem Solver.

C.8 Functions

The functions portion of Problem Solver handles graphing, solving equations, and estimating limits, derivatives, and integrals. Problem Solver handles two types of functions: functions defined by formulas, and functions defined by tables. The tables are made up of two columns. The first is for the x-values, and the second is for the corresponding y-values.

We shall discuss each option available from the functions menu.

```
F1-help
-------------------------------------
Enter/edit functions-- formula or table
Load file
Save file
Remove functions
Variable name
[Functional values]
Sequences
Maxima/minima
Graph Functions
Limits
Derivatives
Integrals
Solve f(x)=0
Options
```

C.8.1 Managing Functions

Enter/Edit Functions: You can enter up to 10 functions for use in any section of this menu. These functions can be entered as formulas or tables of ordered pairs. If you want to enter the function $f(x) = 3x - 4$, enter only the right side as `3*x-4`. If you enter ordered pairs, you will probably want to enter the x-values in increasing order. If you graph your points and elect to have the points connected with line segments, the points will be connected in the order that you enter them.

To graph a multiline function enter the various portions as different functions. Then select [Change graph specifications] to adjust the domain for each portion of your function.

Load and Save: Functions are saved in sets of 10. In addition to the functions themselves, graph specifications are also saved. No special information involving limits, derivatives, integrals, or solving equations is saved.

Remove Functions: Use this option to delete functions. You can remove any single function or all of the functions.

Variable Name: You can change the variable name, but if you do, the former variable will not be recognized. This means that if you change the x to t, x will no longer be recognized as a variable. Using x in an expression will cause an error message to be displayed. If you have already defined functions in terms of x, you must edit them and replace the xs with ts.

C.8.2 Calculations with Functions

Functional Values: This option generates tables of functional values.

```
F1-help
f1(x) = sin(x)
Enter x-values from keyboard
Get x-values from formula
Calculate functional values
Put table in function list
Options
```

After selecting one of your functions, you will use the menu shown above to help you build a table of functional values. You can enter the x-values from the keyboard, or you can use a formula to generate the set of x-values. If you need to enter many values, using a formula will save time. By selecting the next to last option, you can save the table that you generate as one of your list of 10 functions.

```
F1-help
Formula for calculating x-values:
        x[1] = -5
For i=2,3,...,10
        x[i] = -5 + 1*(i-1)
Calculate x-values
```

The menu shown above is used for calculating x-values from a formula. The x-values are defined as the sequence $x_1, x_2, \ldots, x_n$. You must enter the first value; the rest can be calculated from a formula that defines the ith value x_i. Problem Solver uses the notation $x[i]$ for x_i. The formula for $x[i]$ can use three variables: $x[1]$, the first value; $x[i-1]$, the previous value; and i, the index of the current term.

Consider defining the sequence $3, 5, 7, \ldots, 99$. Each x-value is 2 greater than the preceding value. This means that $x[i]$ is 2 greater than $x[i-1]$. We set

$x[1] = 3$, and for $i = 2, 3, \ldots, 99$, we set $x[i] = x[i-1] + 2$. Then $x[1] = 3$, $x[2] = 3 + 2 = 5$, $x[3] = 5 + 2 = 7$, etc. In this formula, each x-value is given as a function of the preceding x-value. Formulas like this are called *recurrence relations.*

There are other ways of producing this sequence of numbers. Consider setting $x[1] = 3$ and $x[i] = 3 + (i-1) \cdot 2$. Then $x[2] = 3 + 1 \cdot 2 = 5$, $x[3] = 3 + 2 \cdot 2 = 7$, etc. In this formula, each x-value is given as a function of the index i.

After the x-values have been entered or calculated, you can request that the functional values be calculated. The results will be displayed in a table. The first column contains the x-values, and the second column contains the functional values. If not all of the values are visible, you can use the cursor control keys to scroll through the table.

Sequences: This option allows you to calculate terms in a sequence. This is done with a formula in exactly the same manner as calculating the x-values of a function table as described above.

Maxima/Minima: Problem Solver estimates the maximum and minimum values of a function defined on $[a, b]$. The program does this by calculating functional values for a set of points in $[a, b]$ and picking the largest and smallest values. By carefully selecting the points, Problem Solver attempts to return values accurate to all of the decimal places displayed.

C.8.3 Graphing Functions

Problem Solver can graph functions entered as formulas or as tables of ordered pairs. Selecting Graph functions brings up the menu shown below:

```
F1-help

Function selection
Change graph specifications
Draw graph
```

Before any functions can be graphed, you must indicate which of the current selection of 10 functions you wish to graph.

```
F1-help
f1(x) = sin(x) yes
f2(x) = cos(x) no
f3(x) = undefined no
.
.
.
f10(x)=undefined no
```

Up to 10 functions can be graphed on the same coordinate system. These functions can be selected individually as shown in the menu above. Any combination of functions can be graphed.

C.8.4 Modifying and Enhancing Graphs

Select [Change graph specifications] to add annotation to your graphs or to adjust the range of values being graphed.

```
F1-help
f1(x) = sin(x)
Main heading:  none
Subheading:  none
Label for x-axis:  none
Label for y-axis:  none
Label for function:  none
Connect points with line segments:  yes
Minimum x-value to graph:  -5
Maximum x-value to graph:  5
Minimum x-coordinate on axis:  -5
Maximum x-coordinate on axis:  5
Minimum y-coordinate on axis:  -3
Maximum y-coordinate on axis:  3
Change min/max values to graph in all functions
```

Function: Use this line to modify the function.

Main Heading: The main heading is centered and displayed above the graph in large type. This heading is the same for all 10 functions.

Subheading: The subheading is displayed just below the main heading and in slightly smaller type.

Label for x-axis: This label appears just below the horizontal axis.

Label for y-axis: The label for the y-axis appears to the right of the y-axis. It is displayed sideways.

Label for Function: The function label is used to identify a graph when several graphs are shown simultaneously. It is not usually used for graphs of single functions.

Connect Points with Line Segments: Problem Solver graphs functions defined by formulas using the highest possible resolution. Still, sometimes a portion of a graph can be so steep that the graph appears as a series of dots. Answer "yes" to this option to connect the points with line segments. Note that if your function is not continuous, it is not appropriate to connect the points. For example, you would not want to connect the points in the graph of the greatest integer function.

You can also connect the points with line segments in a graph of a function defined by a table.

Minimum and Maximum x-values to Graph: These values do not affect the coordinate system. They only serve to limit the graph of the function to an interval.

Minimum and Maximum x-coordinates on Axis: These values change the scale on the coordinate axes. Thus these values affect all 10 functions.

Minimum and Maximum y-coordinates on Axis: These values also affect all 10 functions since they change the scale on the y-axis.

Change Min/Max Values to Graph in All Functions: The minimum and maximum x-values to be graphed can be set independently in each function. However, if you want the settings for one function to apply to all of the functions, this menu option will help you do that.

C.8.5 Limits

Rather than actually calculating limits, Problem Solver helps the user to estimate limits and to study the concepts behind limits. The user must first enter an estimate of the limit, and then Problem Solver will generate a table of values to see whether or not the estimate is reasonable. All of the various types of limits of functions of a single variable can be investigated. This includes right-hand, left-hand, and infinite limits. The following forms of the standard definitions are used:

Definition 16 *(Two-sided limit) The statement*

$$\lim_{x \to a} f(x) = L$$

means that for every $\epsilon > 0$ *there exists a corresponding* $\delta > 0$ *such that*

$$\text{if } 0 < |x - a| < \delta \text{ then } |f(x) - L| < \epsilon.$$

Definition 17 *(Left-hand limit) The statement*

$$\lim_{x \to a^-} f(x) = L$$

means that for every $\epsilon > 0$ *there exists a corresponding* $\delta > 0$ *such that*

$$\text{if } 0 < a - x < \delta \text{ then } |f(x) - L| < \epsilon.$$

Definition 18 *(Right-hand limit) The statement*

$$\lim_{x \to a^+} f(x) = L$$

means that for every $\epsilon > 0$ *there exists a corresponding* $\delta > 0$ *such that*

$$\text{if } 0 < x - a < \delta \text{ then } |f(x) - L| < \epsilon.$$

Definition 19 *(x tends to infinity) The statement*

$$\lim_{x \to \infty} f(x) = L$$

means that for every $\epsilon > 0$ *there exists a corresponding* N *such that*

$$\text{if } x > N \text{ then } |f(x) - L| < \epsilon.$$

Definition 20 *(x tends to negative infinity) The statement*

$$\lim_{x \to -\infty} f(x) = L$$

means that for every $\epsilon > 0$ *there exists a corresponding* N *such that*

$$\text{if } x < N \text{ then } |f(x) - L| < \epsilon.$$

Definition 21 *($f(x)$ tends to infinity) The statement*

$$\lim_{x \to a} f(x) = \infty$$

means that for every M *there exists a corresponding* $\delta > 0$ *such that*

$$\text{if } 0 < |x - a| < \delta \text{ then } f(x) > M.$$

Definition 22 *($f(x)$ tends to negative infinity) The statement*

$$\lim_{x \to a} f(x) = -\infty$$

means that for every M *there exists a corresponding* $\delta > 0$ *such that*

$$\text{if } 0 < |x - a| < \delta \text{ then } f(x) < M.$$

Other limits that can be studied using Problem Solver are

$$\begin{aligned}
\lim_{x\to a^-} f(x) &= \infty \\
\lim_{x\to a^+} f(x) &= \infty \\
\lim_{x\to a^-} f(x) &= -\infty \\
\lim_{x\to a^+} f(x) &= -\infty \\
\lim_{x\to \infty} f(x) &= \infty \\
\lim_{x\to -\infty} f(x) &= \infty \\
\lim_{x\to \infty} f(x) &= -\infty \\
\lim_{x\to -\infty} f(x) &= -\infty
\end{aligned}$$

We now discuss the options available in the limits menu.

```
F1-help
Function:  f(x) = sin(x)
Limit as x approaches 0
User's estimate of limit:  0
Type of limit:  two-sided
Number of lines in table:  20
Compute estimate for δ or N
Graph y = f(x)
Adjust ε or M in the line below.
|f(x) - 0| < 0.5 whenever
0 < |x - 0| < δ
```

Two things must be done in order to examine limits with Problem Solver:

1. The user must enter an estimate for the limit. The graphing option can be used to help in this step.

2. A value of δ or N must be calculated.

Function: Select this option if you want to change the function.

Limit as x approaches a: Use this option to change the value of a to any real number, infinity, or negative infinity.

User's Estimate of Limit: Use this value to change your estimate of the limit to any real number, infinity, or negative infinity.

Type of Limit: The type of limit can be two sided, left-hand, or right-hand.

Number of Lines in Table: Problem Solver checks limits by calculating tables of functional values. You may select the number of lines in the table.

Compute Estimate for δ or N: The menu shown below gives the specifications for the table used by Problem Solver to estimate δ from the given value of ϵ.

```
F1-help

Formula for calculating x-values:
        x[1] = 1
For i=2,3,...,20
        x[i] = 0 + (-.5)^(i-1)
Calculate x-values
```

The x-values are calculated from a formula. These values should be approaching the number a that x is approaching in the limit. As an example, consider

$$\lim_{x \to 4} 2x = 8.$$

The first step in examining this limit is to choose a number ϵ. Problem Solver will estimate a value of δ such that

$$\text{if } 0 < |x - 4| < \delta \text{ then } |2x - 8| < \epsilon.$$

Suppose that we select $\epsilon = 0.3$. Problem Solver uses a table of functional values to estimate δ. To construct this table, we define a sequence of x-values approaching 4. We could set `x[1]=5`, and `x[i] = 4 + (-1/2)^(i-1)`. These are the default settings for this problem. Then

```
x[1] = 4 + (-1/2)^0 = 5.00000
x[2] = 4 + (-1/2)^1 = 3.50000
x[3] = 4 + (-1/2)^2 = 4.25000
x[4] = 4 + (-1/2)^3 = 3.87500
x[5] = 4 + (-1/2)^4 = 4.06250
x[6] = 4 + (-1/2)^5 = 3.96875
```

The function table follows:

x	$f(x)$
5.00000	10.00000
3.50000	7.00000
4.25000	8.50000
3.87500	7.75000
4.06250	8.12500
3.96875	7.93750

Problem Solver will examine the table and conclude that $|2x-8| < 0.3$ for the last three lines of the table. That is,

$$\text{if } 0 < |x-4| < 0.125 \text{ then } |2x-8| < 0.3,$$

and $\delta = 0.125$. It can be shown analytically that we could have selected δ as large as 0.15. That is,

$$\text{if } 0 < |x-4| < 0.15 \text{ then } |2x-8| < 0.3.$$

As the example above shows, selecting δ from a table doesn't usually give the largest possible value. Looking at a different table may suggest a different value of δ.

If you want the x-values to approach 4 more rapidly, replace 1/2 in the formula with a smaller positive number. For a more leisurely approach, replace 1/2 with a number closer to but still less than 1. In this case you may need to increase the number of lines in the function table.

If you want these numbers to approach 4 from the right, replace $-1/2$ with 1/2. For an approach from the left, set `x[i] = 4 - (1/2)^(i-1)`.

Graph $y = f(x)$: After an appropriate value of δ (or N if x tends to infinity) has been calculated, Problem Solver can illustrate the relationship of ϵ to δ (or M to N) graphically.

Adjust ϵ or M in the Line Below: To adjust the current value of ϵ, move the cursor to the next line, and press the Enter key.

C.8.6 Derivatives

Problem Solver estimates first and second order derivatives for a finite set of x-values. The derivative estimates can be presented as a table of values or as a graph.

```
F1-help
```

```
Function:  f(x) = sin(x)
Highest order derivative:  1
Compute table
Draw graph 5
Explore secant and tangent lines
```

Function: Use this option to change the function.

Highest Order Derivative: First and second order derivatives can be studied.

Compute Table: Problem Solver approximates derivatives by calculating tables of values. You must define a sequence $x_0, x_1, \ldots, x_n$ of x-values. If these numbers are evenly spaced, h is set equal to the common distance between consecutive x- values. In this case, Problem Solver begins by using the following formulas with $f_i = f(x_i)$.

$$\begin{aligned}
f'(x_i) &= \frac{-3f_i + 4f_{i+1} - f_{i+2}}{2h} \\
f''(x_i) &= \frac{2f_i - 5f_{i+1} + 4f_{i+2} - f_{i+3}}{h^2} \\
f'''(x_i) &= \frac{-5f_i + 18f_{i+1} - 24f_{i+2} + 14f_{i+3} - 3f_{i+4}}{2h^3} \\
f^{(iv)}(x_i) &= \frac{3f_i - 14f_{i+1} + 26f_{i+2} - 24f_{i+3} + 11f_{i+4} - 2f_{i+5}}{h^4}.
\end{aligned}$$

Problem Solver begins with $i = 0$ and continues to $i = 1$ if necessary so that just enough values are calculated to allow it to jump to the next group of formulas. For first and second order derivatives, it needs only one value from the first group of formulas, but for third and fourth order derivatives, it needs two values.

The second group of formulas provide more accurate results, so Problem Solver uses them as much as possible.

$$\begin{aligned}
f'(x_i) &= \frac{-f_{i-1} + f_{i+1}}{2h} \\
f''(x_i) &= \frac{f_{i-1} - 2f_i + f_{i+1}}{h^2} \\
f'''(x_i) &= \frac{-f_{i-2} + 2f_{i-1} - 2f_{i+1} + f_{i+3}}{2h^3} \\
f^{(iv)}(x_i) &= \frac{f_{i-2} - 4f_{i-1} + 6f_i - 4f_{i+1} + f_{i+2}}{h^4}.
\end{aligned}$$

Since the second group of formulas requires x-values both smaller and larger than x_i, Problem Solver can not use these formulas near the end of the table. When necessary, it switches to the third and last group of formulas:

$$\begin{aligned}
f'(x_n) &= \frac{-3f_n + 4f_{n-1} - f_{n-2}}{2h} \\
f''(x_n) &= \frac{2f_n - 5f_{n-1} + 4f_{n-2} - f_{n-3}}{h^2} \\
f'''(x_n) &= \frac{-5f_n + 18f_{n-1} - 24f_{n-2} + 14f_{n-3} - 3f_4}{2h^3} \\
f^{(iv)}(x_n) &= \frac{3f_n - 14f_{n-1} + 26f_{n-2} - 24f_{n-3} + 11f_{n-4} - 2f_{n-5}}{h^4}.
\end{aligned}$$

If your x-values are not evenly spaced, only the first derivative can be estimated, and the following formulas are used:

$$f'(x_i) = \frac{f_{i+1} - f_i}{h} \text{ for } i = 0, 1, ..., n-1$$
$$f'(x_n) = \frac{f_n - f_{n-1}}{h}.$$

For many functions and evenly spaced data, the error is approximately proportional to the square of h. This means that if you use twice as many x-values in your estimate so that h is divided by 2, you will divide the error by approximately 4.

If you want to estimate a derivative at a single point, create a table of values with your point near the center of the table. From the formulas above you can see that you need at least 3 lines in your table.

Draw Graph: Graphs of first and second order derivatives can be displayed. These derivatives are calculated as a set of ordered pairs, which are graphed and connected with line segments.

Explore Secant and Tangent Lines: The user can view a sequence of secant lines approaching the tangent line at a point $(a, f(a))$. The ith secant line passes through the points $(a, f(a))$ and $(x_i, f(x_i))$. The user controls both the number of secant lines and the rate at which x_i approaches a.

Third and fourth order derivatives are used internally by Problem Solver to calculate error estimates for the integration methods. They are not directly available to the user.

C.8.7 Integrals

Definite integrals of functions defined either as formulas or tables with uniformly spaced x-values can be estimated using Riemann sums, the trapezoidal rule, or Simpson's rule. For functions defined as tables of values with unevenly spaced x-values, only the method of Riemann sums can be used.

```
F1-help

Integrand:  sin(x)
Lower limit of integration:  0
Upper limit of integration:  1
Algorithm:  Simpson's rule
Number of subintervals:  30
Draw graph
Compute integral
Options
```

Integrand: Use this option to change the function.

Lower and Upper Limit: These options allow you to change the limits of integration.

Algorithm: Select from the trapezoidal rule, Simpson's rule, or Riemann sums. In the case of Riemann sums, you can choose to have the function evaluations done at one of the endpoints of the subintervals or at the midpoints of the subintervals. Simpson's rule ordinarily provides the most accurate results.

Number of Subintervals: For Simpson's rule, select an even number of subintervals. For functions defined as tables of values, the method of Riemann sums with the function evaluation done at the midpoints also requires an even number of subintervals.

Draw Graph: The function being integrated can be graphed. If $a < b$, the area bounded by the graph of $y = f(x)$, the lines $x = a$ and $x = b$, and the x-axis is shaded. If $f(x) \geq 0$ for $x \in [a, b]$, the value of the integral is equal to this area.

Compute Integral: For functions defined as formulas on the interval $[a, b]$, Problem Solver first subdivides $[a, b]$ into n subintervals of equal length by defining

$$a = x_0 < x_1 < \cdots < x_n = b$$

where $h = (b - a)/n = x_i - x_{i-1}$ for $i = 1, 2, \ldots, n$.

For Riemann sums, the user selects $w_i \in [x_{i-1}, x_i]$ as the midpoint or one of the endpoints. Then the formula

$$h(f(w_1) + f(w_2) + \cdots + f(w_n))$$

is used to estimate the integral. The other formulas are

$$\frac{h}{2}(f_0 + 2f_1 + 2f_2 + \cdots + 2f_{n-1} + f_n)$$

and

$$\frac{h}{3}(f_0 + 4f_1 + 2f_2 + 4f_3 + 2f_4 \cdots + 2f_{n-2} + 4f_{n-1} + f_n),$$

for the trapezoidal rule and Simpson's rule, respectively, where $f_i = f(x_i)$ for $i = 1, 2, \ldots, n$.

For functions defined as tables of values, the x_is are the x-values in the function table. Functional values at other x-values are not available. This presents a problem if Riemann sums are used with w_i selected as the midpoint of the interval $[x_{i-1}, x_i]$. Since it is impossible to compute a functional value at the midpoint of this interval, Problem Solver uses the interval $[x_{i-1}, x_{i+1}]$ with midpoint x_i. That is, the subintervals are used two at a time. This means that an even number of subintervals is required. Thus for functions defined by tables of ordered pairs, an even number of subintervals is required for both Simpson's rule and Riemann sums with w_i chosen as the midpoint of $[x_{i-1}, x_i]$.

Options: The options menu is also available from the main menu. See Section C.7.

Error estimates are calculated for integrals. If I^* is the estimate for the actual value I of an integral, the error is $|I - I^*|$, and the relative error is

$$\frac{|I - I^*|}{|I|}.$$

Problem Solver uses standard formulas based on derivatives to estimate an upper bound or worst-case estimate for the error. Since I is not known, Problem Solver estimates the relative error bound using

$$\frac{|I - I^*|}{|I^*|}.$$

C.8.8 Solving Equations

Equations of the form $f(x) = 0$ can be solved using the bisection method, the secant method, or Newton's method.

```
F1-help

Function:  sin(x)
Derivative:  not used
Initial guess:  0 1
Method:  Bisection method
Number of iterations:  50
Maximum error:  5e-06
Maximum residual error:  5e-10
Compute solution
Options
```

Function: Use this option to enter a new function. Note that to solve the equation $f(x) = 0$, one enters $f(x)$ on this line. The equal sign and the zero are not entered.

Derivative: Newton's method uses the derivative of $f(x)$. Enter it on this line. This derivative can also be entered from the functions menu by selecting Enter/edit functions. In this case, the derivative of $f(x)$ should be entered in the line below $f(x)$.

Initial Guess: The methods used need at least one initial guess. These numbers should be your best estimates of the solution. It might be helpful to draw graphs of these functions and estimate the x-intercepts of the graphs. See the explanation of the methods for more details on selecting initial guesses.

Method: Newton's method, the secant method, and the bisection method can be used. In each of these methods, a sequence of approximate solutions is calculated.

For Newton's method, x_0 is an initial guess, and

$$x_{i+1} = x_i - \frac{f(x)}{f'(x)}$$

for $i = 0, 1, \ldots$. With a good first guess, Newton's method usually gives very good results in just a few steps. For many functions, the number of correct decimal places in the solution approximately doubles in each step.

The secant method is nearly as fast as Newton's method. It differs from Newton's method in that it uses an approximation for the derivative rather than the actual derivative. To get started, the secant method requires two guesses: x_0 and x_1.

Then

$$x_{i+2} = x_{i+1} - \frac{f(x_{i+1})(x_{i+1} - x_i)}{f(x_{i+1}) - f(x_i)}$$

for $i = 0, 1, \ldots$.

The bisection method also requires two initial guesses: x_0 and x_1. In addition, the function must be continuous on $[x_0, x_1]$, and the functional values $f(x_0)$ and $f(x_1)$ must differ in sign.

First the midpoint of the interval $x_m = (x_0 + x_1)/2$ is found, and $f(x_m)$ is calculated. If $f(x_0)$ and $f(x_m)$ differ in sign, x_1 is replaced with x_m, and then we start over. If $f(x_m)$ and $f(x_1)$ differ in sign, x_0 is replaced with x_m, and we begin again.

While both Newton's method and the secant method can fail in some cases, the bisection method always works. However, for a given level of accuracy, the bisection method usually requires far more steps than either of the other two methods.

Number of Iterations: This is the maximum allowable number of iterations.

Maximum Error: If x^* is the approximate solution and x is the actual solution, the error is $|x - x^*|$. As long as this error tolerance is not met, Problem Solver will keep iterating (up to the maximum number of iterations).

Maximum Residual Error: If x^* is the approximate solution, $|f(x^*)|$ is the residual error. Problem Solver will continue its calculations up to the maximum number of iterations trying to meet this error tolerance.

Compute Solution: Use this option to find an approximate solution.

Options: This menu is available from the main menu too. See Section C.7 for more details.

C.9 Matrices

The matrix options include solving systems of equations, finding sums and products of matrices, finding inverses of matrices, row reducing matrices, and solving linear programming problems.

```
F1-help

Linear equations/inverses
Matrix operations
Linear programming
```

Both matrices and linear programming models are entered and edited with the matrix editor. This screen consists of a rectangular array of cells for your matrix entries. You can enter or edit an entry by moving the cursor to that entry, and typing the new entry. If you have ever used a spreadsheet program, you will immediately feel comfortable with this screen. The following commands are available from this editor:

Right Arrow: Move cursor one cell to the right.

Left Arrow: Move cursor one cell to the left.

Up Arrow: Move the cursor up one cell.

Down Arrow: Move the cursor down one cell.

Home: Move the cursor to the top left cell.

Ctrl-Right Arrow: Move the cursor to the right one screen.

Ctrl-Left Arrow: Move the cursor to the left one screen.

PgUp: Move the cursor up one screen.

PgDn: Move the cursor down one screen.

F2: Edit the current entry.

F3: Load a data file directly into the editor.

F4: Save the current data file.

F5: Change the dimensions of your problem. If you are using this editor for data for running descriptive statistics, all of the data must be put in one column. You will not be allowed to increase the number of columns. For data for linear regression, you must use exactly two columns. The first is for the independent variable (x-values), and the second is for the dependent variable (y-values).

F6: Use this to begin with a new matrix.

F7: The cursor may be set to move ahead automatically after an entry is made. The direction of the automatic move is the same as the last manual move.

Del: Delete an entry, row, or column.

Ins: Insert an entry, row, or column.

Esc: Leave the editor.

Deleting Entries, Rows, or Columns

Entries: First move the cursor to the entry that you wish to delete. Then press the Del key. When the entry is deleted, the entries to the right will be shifted to the left, and the extreme right entry will be set to zero.

Rows: Move the cursor to the row that you wish to delete. Then press the Del key. This option reduces the number of rows by one. In a linear programming problem, the first row represents the objective function, and it can not be deleted.

Columns: First move the cursor to the column that you wish to to delete. Then press the Del key. This option reduces the number of columns by one. In a linear programming problem, neither of the last two columns can be deleted.

Inserting Entries, Rows, or Columns

Entries: Move the cursor to the position of the entry you wish to insert and press the Ins key. The entries to the right will be shifted to the right and the new entry will be placed at the cursor position and set to zero.

Rows: Move the cursor to the position of the new row, and then press the Ins key. The new row will have zero in each entry, and all rows below will be shifted down. The number of rows will be increased by one.

Columns: Move the cursor to the position of the new column, and then press the Ins key. The new column will have zero in each entry, and all columns to the right will be shifted to the right. The number of columns will be increased by one.

C.9.1 Linear Equations/Inverses

```
F1-help
-------------------------
Enter/edit matrix
Load matrix
Save matrix
Remove matrix
[Compute solution/inverse]
Options
```

Enter/Edit Matrix: This option allows you to use the matrix editor described at the beginning of this section. To solve a system of equations, enter the augmented matrix associated with the system.

The augmented matrix for the system of equations

$$\begin{array}{rcrcrcr} 2x & - & 3y & + & 5z & = & 9 \\ 3x & & & + & 6z & = & 10 \\ x & + & 8y & - & z & = & 11 \end{array}$$

is

$$\begin{pmatrix} 2 & -3 & 5 & 9 \\ 3 & 0 & 6 & 10 \\ 1 & 8 & -1 & 11 \end{pmatrix}.$$

This matrix has 3 rows and 4 columns, and it is the matrix that should be entered into Problem Solver.

Load/Save Matrix: Matrices can be saved in files for future use.

Remove Matrix: Removing a matrix from memory frees memory for other uses.

Compute Solution/Inverse: This option allows you to find solutions of systems of equations, inverses of matrices, and row-reduced forms of matrices.

```
F1-help
-------------------------
Gaussian elimination
Gauss-Jordan method
Inverse matrix
[Reduced-row echelon form]
```

Gaussian Elimination: This option uses Gaussian elimination with partial pivoting and backward substitution.

Gauss-Jordan Method: The Gauss-Jordan method is used with partial pivoting.

Inverse Matrix: To find the inverse of a square matrix A, Problem Solver augments A with the identity matrix and converts this new matrix to row-reduced echelon form. If the left half is the identity matrix, the right half is the inverse. This means that for Problem Solver to find the inverse of an $n \times n$ matrix, enough memory must be available for the $n \times 2n$ augmented matrix.

Reduced Row-Echelon Form: Use this option to convert any matrix to reduced form. This is useful for studying systems whose coefficient matrices are singular. By studying the reduced form you can tell whether the system has multiple solutions or no solutions at all.

C.9.2 Matrix Operations

```
F1-help

Enter/edit matrix
Load matrix
Save matrix
Remove matrix
Print matrix
SUM ← A + B
PROD ← A*B
PROD ← A*PROD
A ← inverse of A
A ← transpose of A
A ← kA
A ← PROD
A ← SUM
A ← B
A ↔ B
User controlled matrix reduction
Options
```

Four matrices are available for the matrix operations. These are named A, B, SUM, and $PROD$. The matrix A is the same as the matrix used in the linear equations/inverses menu. The others are used only in the matrix operations menu.

The notation "$SUM \leftarrow A+B$" means that the sum of A and B is to be calculated and placed in the matrix called SUM. Similarly the notation "$A \leftarrow$ inverse of A" means that the inverse of A is to be found and that this inverse is placed in the matrix A. We now consider the matrix operations menu options.

Enter/Edit Matrix: Any of the matrices A, B, SUM, or $PROD$ can be entered or edited.

Save/Load Matrix: Any of the matrices A, B, SUM, or $PROD$ can be saved for later use.

Remove Matrix: You can remove any of the matrices A, B, SUM, or $PROD$ from memory, thereby making the memory available for other uses.

Print Matrix: Select this option if you want to print your matrix.

SUM ← A + B: The sum of the matrices A and B is calculated, and this sum is placed in the matrix SUM.

PROD ← A*B: The product of the matrices A and B is calculated, and this product is placed in the matrix $PROD$.

PROD ← A*PROD: The product of the matrices A and $PROD$ is calculated, and this product is placed in the matrix $PROD$.

A ← Inverse of A: The inverse of the matrix A is calculated and placed in the matrix A. This means that the original matrix A is no longer available. Selecting this option a second time will restore the matrix A to its original condition.

A ← Transpose of A: The transpose of A is calculated and placed in the matrix A. This means that the rows and columns of A are interchanged.

A ← kA: The matrix A is multiplied by the real number k. This product becomes the new matrix A.

A ← PROD: The matrix $PROD$ is renamed A, and the old matrix A is removed.

A ← SUM: The matrix SUM is renamed A, and the old matrix A is removed.

A ← B: The matrix B is renamed A, and the old matrix A is removed.

A ↔ B: The matrix A is renamed B, and the matrix B is renamed A.

User Controlled Matrix Reduction: Select this option if you want to control the row operations in the reduction of a matrix. The following options are available:

```
F1-help
```

```
First row index:  1
Second row index:  2
Interchange rows 1 and 2
Multiplier:  4
Row 1 ← 4*Row 1
Row 2 ← Row 2 + 4*Row 1
Print matrix
Options
```

First row index: Enter the index of the row being used to modify other rows. This row is called the *pivot row*. To reduce column j, use row j as the pivot row.

Second row index: Enter the index of the row being modified.

Interchange rows: Use this option to interchange the two rows identified in the first two options.

Mutliplier: Enter a number here to be used in the next two options.

Row 1 ← 4*Row 1: The first row identified above will be multiplied by the number entered into the above option. In this illustration 4 is the multiplier.

Row 2 ← Row 2 + 4*Row 1: This is the main step used in matrix reduction. The row selected in the second line of this menu is replaced by the sum of itself and the product of the multiplier and the row selected in the first line.

Print matrix: Select this option to print the current version of your matrix.

Options: This is the same options menu that is available from the main menu.

C.9.3 Linear Programming

```
F1-help
```

```
Enter/Edit model
Load model
Save model
Remove model
Objective:  maximize
Compute optimal solution
Options
```

Enter/Edit Model: Linear programming models are entered and edited using the matrix editor. In the linear programming mode, the first line of the edit screen is reserved for the objective function, and the next to last column is reserved for the relational operators. Otherwise this editor works with linear models in exactly the same way it works with matrices. As an example, consider the following linear programming problem: Find the maximum value of

$$Z = x_1 + 2x_2$$

subject to the constraints

$$\begin{array}{rcrcl} 3x_1 & + & x_2 & \leq & 4 \\ x_1 & + & 2x_2 & \leq & 3 \\ x_1 & + & x_2 & \leq & 6 \end{array}$$

$$x_1 \geq 0, x_2 \geq 0.$$

After this problem has been entered into Problem Solver, the edit screen should look like this:

	1	2	3	4
obj	1	2		
1	3	1	<	4
2	1	2	<	3
3	1	1	<	6

Load/Save Model: Linear programming models can be saved for later use.

Remove Model: Removing a model frees memory, which is then available for other uses.

Objective: The objective function can be either maximized or minimized.

Compute Solution: Use this option to find the optimal solution for the linear programming model. During the solution process the ordering of the constraints may be changed.

The options menu allows you to display the intermediate calculations used to find the optimal solution. Problem Solver uses a matrix form of the simplex algorithm called the *two-phase revised simplex algorithm.* However, the intermediate steps are displayed in the tableau form shown below:

	1	2	3	4	5	6	7
obj	−1	−2	0	0	0		
x3	3	1	1	0	0	=	4
x4	1	2	0	1	0	=	3
x5	1	1	0	0	1	=	6

The current basic feasible solution is also displayed. In the tableau, the variables are identified by their subscripts in the top row. The left column identifies the current basic variables, and the right column gives the current values of the basic variables. These numbers are color coded according to their type: decision, slack, surplus, or artificial. In addition, the row and column containing the entering and leaving variables are highlighted. If the tableau is too large to fit on the screen, the screen acts as a window to allow you to view a portion of the tableau. Use the cursor control keys to see other portions of the tableau.

Because showing intermediate calculations requires more memory than simply solving the problem, some models that are too large to be solved if the intermediate calculations are displayed may be manageable if no intermediate calculations are shown. For more details on the Simplex method, see Appendix D.

Options: The options screen is also accessed from the main menu. See Section C.7 for more detail.

C.10 Statistics

Problem Solver handles probability problems, counting problems, descriptive statistics including histograms, and linear regression.

```
F1-help

Probability/counting
Descriptive statistics
Linear regression
```

We shall discuss each of these options.

C.10.1 Probability and Counting

Select Probability/counting if you want to calculate factorials, permutations, combinations, or probabilities using the normal or binomial distribution.

```
F1-help

Permutations/combinations
Binomial distribution
Normal distribution
Options
```

Permutations/Combinations: The permutations/combinations option allows you to calculate factorials, count ordered arrangements of objects, and count the number of subsets of a given size from a set of objects.

```
F1-help
Total number of objects:   n = 10
Number of selected objects:   r = 21
Factorial:  n!
Permutations:  P(n,r)
Combinations:  C(n,r)
Options
```

This menu is one of two ways that Problem Solver handles permutations and combinations. Problem Solver provides the function $P(n, r)$, which calculates the number of permutations of n objects taken r at a time, and $C(n, r)$, which calculates the number of combinations of n objects taken r at a time. The function $fact(n)$ gives the value of $n!$. These functions can be used in the calculator provided in the options menu.

The calculator method is particularly useful for probability problems. For example, one could use the calculator to evaluate C(3,2)*C(4,2)/C(7,4). This is considerably more efficient than calculating one term at a time using the menus, and then doing the multiplication and division separately on a calculator.

Total Number of Objects n: This number should be a positive integer.

Number of Selected Objects r: This number is an integer between 0 and n inclusive.

Factorial $n!$: This is the product $1 \cdot 2 \cdot 3 \cdots n$.

Permutations $P(n, r)$: This gives the number of ordered arrangements of n objects taken r at a time. It is calculated from the formula

$$P(n,r) = \frac{n!}{(n-r)!}.$$

Combinations $C(n, r)$: This gives the number of subsets of size r in a set of size n. It is calculated from the formula

$$C(n,r) = \frac{n!}{r!(n-r)!}.$$

Options: See Section C.7.

Binomial Distribution: Select this option to calculate probabilities using the binomial distribution.

```
F1-help
Total number of trials:  n = 10
Probability of success in each trial:  p = 0.5
Lower limit:  a = 2
Upper limit:  b = 4
Binomial probability:  P(a≤X≤ b)
Options
```

Total Number of Trials: This positive integer gives the number of trials that comprise the binomial experiment.

Probability of Success in Each Trial: This number should be between 0 and 1.

Lower Limit a and Upper Limit b: These values are used in inequality in next menu line.

Binomial Probability: Select this item to calculate the probability that the binomial random variable X has one of the values $a, a+1, a+2, \ldots, b$. To calculate the probability that $X = c$, give both a and b the value c.

Normal Distribution: These probabilities are calculated using the normal distribution.

```
F1-help
Mean:  μ = 0
Standard deviation:  σ = 1
Lower limit:  -0.5
Upper limit:  0.4
Draw graph
Normal probability:  P(a ≤ X ≤ b)
Options
```

Mean and Standard Deviation: These parameters are used to specify a particular normal distribution.

Lower Limit a and Upper Limit b: Input a and b to find $P(a \leq X \leq b)$. To calculate $P(a \leq X)$, enter a large number for b. Five standard deviations above the mean should be adequate. That is, set $b = \mu + 5\sigma$. To calculate $P(X \leq b)$, set $a = \mu - 5\sigma$. Since expressions can be entered for numbers, you can let Problem Solver do the calculations. For example, suppose that $\mu = 2$, and $\sigma = 5$. To calculate $P(X \leq 6)$, enter 6 for b and `2-5*5` for a.

Draw Graph: This option allows you to view the graph of the normal curve with a shaded region corresponding to the probability calculation. The shaded area is the probability $P(a \leq X \leq b)$.

Normal Probability: Select this option to calculate that X has a value between a and b.

C.10.2 Descriptive Statistics

```
F1-help
-------------------------------------
Enter/edit data
Load data
Save data
Remove data matrix
Print data matrix
[Generate random data]
Mean/median/standard deviation
Frequency table
Histogram
Options
```

Enter/Edit Data: The matrix editor is used to enter all data for the statistics routines. For the descriptive statistics procedures, only one column is used for your data. All numbers must be entered into a matrix with a single column.

Load/Save Data: Data can be saved for later use.

Remove Data Matrix: If you are running short of memory, and if you no longer need your data , removing the data matrix will free memory for other uses.

Print Data Matrix: Use this option for printing your data.

Generate Random Data: Random data sets from uniform, binomial, and normal distributions can be generated. These data sets can be used in simulation problems. For each distribution, the user enters parameters and Problem Solver generates a data set of randomly selected points from that distribution.

Mean/Median/Variance/Standard Deviation: Select this option for sample statistics for your data set.

Frequency Table: Frequency tables can be generated to help organize large data sets.

```
F1-help
-------------------------------------
[Change table specifications]
Compute table
```

Change Table Specifications: Problem Solver picks class intervals for your data, but you may wish to control these choices yourself.

```
F1-help
Number of class intervals:  7
Width of class intervals:  3
Left boundary of first interval:  -0.5
Options
```

Number of Class Intervals: The data will be divided into categories or class intervals. You may adjust the number of these categories.

Width of Class Intervals: Enter the width of each interval. The width of the interval $[a, b]$ is $b - a$.

Left Boundary of First Interval: If the intervals have the form $[x_0, x_1], [x_1, x_2], \ldots, [x_{n-1}, x_n]$, then x_0 is the left boundary of the first interval.

Compute Table: After the specifications have all been set, select this option to construct the table.

Histogram: This graphical representation of a frequency table is useful for organizing data.

```
F1-help
Change graph specifications
Graph histogram
```

Change Graph Specifications: The options here are identical to those available under the functions menu. However, the only parameters that should be changed are the minimum and maximum x- and y-values on the coordinate axes.

Graph Histogram: Use this option to display the histogram.

C.10.3 Linear Regression

Linear regression is used for finding a line that best fits a set of ordered pairs.

```
F1-help
Enter/edit data
Load data
Save data
Remove data matrix
Compute regression line
Graph regression line Options
```

Enter/Edit Data: The matrix editor is used to enter data for linear regression. This data is entered into a matrix with two columns.

The x-values or values of the independent variable are entered into the first column, and the y-values or the values of the dependent variable are entered into the second column.

Load/Save Data: Data can be saved for future use.

Remove Data Matrix: Removing a data matrix frees memory for other uses.

Compute Regression Line: Given a set of data points of the form (x_i, y_i) for $i = 1, 2, \ldots, n$, Problem Solver finds the regression line with equation $y = ax+b$ where a and b are selected to minimize the expression

$$\sum_{i=1}^{n}(ax_i + b - y_i)^2.$$

Graph Regression Line: This graph shows both the data points and the regression line.

```
F1-help
Change graph specifications
Graph regression line
```

Change Graph Specifications: The options here are the same as those available under the graph functions menu. However, the only parameters that should be changed are the minimum and maximum x- and y-values on the coordinate axes.

Graph Regression Line: Select this option to see the graph of the data points and the regression line.

C.11 Warnings and Error Messages

C.11.1 Calculation Warnings and Error Messages

- Error. *Division by zero attempted.* Problem Solver encountered a denominator whose absolute value is less than 10^{-100}.

- Error. *Not enough points for derivative estimate.* Second order derivatives require at least three ordered pairs. The fourth order derivatives used for estimating errors in the Simpson's rule for integration require at least 6 ordered pairs. Thus at least 6 subintervals must be used in order to get error estimates.

- Error. *Overflow error.* The value of an expression would have resulted in a number larger than 10^{100}.

- Error. *Values in first column must not be all equal.* This message is given if linear regression is attempted with a data set where all values given for the independent variable are equal.

- Warning. *Artificial variables remain basic at the end of phase 1.* Occasionally a linear programming problem leads a complication in moving from phase 1 to phase 2 of the simplex algorithm. One possible cause of this is redundant constraints. If this is the problem, it may be necessary for the program to remove the redundant constraints. Problem Solver does this by changing all of the coefficients in the constraint to 0.

- Warning. *Error tolerance not met. Solution may be inaccurate.* The selected equation solving routine was not able to meet the required error tolerance in the given number of iterations. Try graphing the function to get improved initial guesses.

C.11.2 File Warnings and Error Messages

- Error. *Can not find help system file. Place the help disk in the disk drive.* The help files are not available to Problem Solver. To access the help system, place the help disk in the default drive, and press a key. To abort your request for help, press the Esc key.

- Error. *Can not open file xxxx.* For some reason, the requested file can not be opened. Most likely you are using an illegal file name. File names should consist of up to 8 letters or digits with an optional suffix of a period and up to 3 more letters or digits. A few characters other than letters or digits can be used. See your DOS manual for more details.

- Error. *Can not open file to save options.* Problem Solver can not open a configuration file with your selected options. The disk may be write protected or already full.

- Error. *File of type xxxx rather than the requested type xxxx.* Problem Solver uses several different types of data files. One type stores functions, a second type stores matrices, and a third type stores linear programming problems. To load a file of a certain type, you must use the menu appropriate for that type of file. You can not load a linear programming file from the load command in the functions menu.

- Error. *File xxxx is not a Problem Solver data file.* You attempted to load a file that is not in the correct format for Problem Solver.

- Error. *Unknown Error.* Problem Solver is unable to determine the cause of this error.

- Warning. *File xxxx exists. Overwrite?* While trying to save a file, you entered a file name that is already in use by another file. If you respond by typing `Y`, the existing file will be replaced by the file you are saving. If you respond by typing `N`, you will be able to save your work under a new file name.

C.11.3 Function and Expression Warnings and Error Messages

- Error. *Attempted to evaluate a blank expression.* An attempt was made to evaluate an expression all of whose characters had been deleted.

- Error. *Begin parenthesis expected.* A left parenthesis is required. Possibly the argument of a function is not enclosed in parentheses.

- Error. *End parenthesis expected.* A left parenthesis was encountered with no matching right parenthesis.

- Error. *Error encountered in the integrand.* While attempting to estimate an integral, Problem Solver was unable to evaluate the integrand at some point. If you request to see the intermediate steps, you will be able to see exactly where the error occurred.

- Error. *Error in function evaluation.* While calculating a function table for studying limits, Problem Solver could not evaluate the function at some of the given x-values. Check for syntax or domain errors.

- Error. *Incorrect function type. The function must be defined by a formula rather than a table.* Certain operations, such as examining limits, require functions defined by formulas. These operations are not available for functions defined by tables.

- Error. *Insufficient memory.* This is a serious problem. Problem Solver is unable to evaluate an expression due to a shortage of memory in your computer. If you are using memory resident programs, remove them. If you have large matrices or other data sets currently in Problem Solver's memory, remove them.

- Error. *No table has been entered.* An attempt was made to use a function table that had not been loaded or entered from the keyboard.

- Error. *Noninteger exponent applied to negative or zero base.* The expression a^b, where b is not an integer, is evaluated as $\exp b \ln a$. This means that a must be positive.

- Error. *Syntax error.* An error was encountered when Problem Solver tried to evaluate an expression that you entered. Problem Solver attempts to indicate the approximate location of the error. Check for typing errors.

- Error. *The function xxxx is not defined at xxxx.* A domain error was encountered for your function. For example, Problem Solver will give this error message if it attempts to evaluate the square root of a negative number.

- Error. *The function xxxx is too large at xxxx.* The value of a function would exceed 10^{100} if the evaluation were carried out.

- Error. *The function xxxx is too near 0 at xxxx.* An attempt was made to evaluate a function at a number so near zero that the evaluation would have led to a result smaller than 10^{-100} or larger than 10^{100}.

- Error. *Two or more ordered pairs have the same first element.* Functions defined as tables of values or sets of ordered pairs must contain no two ordered pairs with the same first element. Each first element must appear only once in the list of ordered pairs.

- Error. *The value of xxxx is not accurate at xxxx.* The functional value is not dependable. You might get this error if you tried to evaluate $sin(10^{99})$.

- Error. *The value passed to xxxx is not defined.* A function argument is not defined.

C.11.4 Graphics Warnings and Error Messages

- Error. *Inadequate screen resolution for requested graph.* Some of the graphs of limits require labels that are too long for the 320 × 200 CGA color mode. Problem Solver always defaults to a graphics resolution of 640 × 200 or better. However, you can reduce the resolution to 320 × 200 by using the command line parameter CGALO. If you are using a CGA system, you might like to use the low resolution mode occasionally since it does add color. This low resolution mode should not be used in the limit section of Problem Solver.

- Error. *No points were plotted for one of the functions.* Problem Solver ordinarily does not report domain errors for functions while it is graphing. It just skips over points that lead to errors. Consequently, you can ask to graph $\sqrt{x}$ for $x \in [-2,2]$ and not worry about the negative numbers. Points are plotted only for $x \in [0,2]$. However, if no points are plotted for one or more of your functions, something is probably wrong. It may be that your function has a syntax error, or it may be that none of the numbers in the given interval are included in the domain of the function. This would be the case if you tried to graph $y = \sqrt{x}$ for $x \in [-10,-5]$. Since all of the numbers in that interval are negative, y would not be defined at any of them, and there would be no graph.

- Error. *Unable to enter graphics mode.* Problem Solver was unable to locate a compatible graphics adapter in your computer. If you have a graphics adapter compatible with the CGA, use the command line parameter CGAHI or CGALO when you start Problem Solver.

C.11.5 Input Warnings and Error Messages

- Error. *Enter or calculate x-values first.* Before a function table can be displayed, the x-values must be entered from the keyboard or computed from a formula.

- Error. *Even number of intervals needed for requested method – switching to the trapezoidal rule.* Simpson's rule requires an even number of intervals. For functions defined as tables of values, the method of Riemann sums using the midpoints of the subintervals also requires an even number of subintervals (an odd number of points).

- Error. *First select functions.* Before graphing functions, you must indicate which of the current collection of 10 functions is to be graphed.

- Error. *Function must change sign at initial guesses.* If the bisection method is started with initial guesses x_1 and x_2, $f(x_1)$ and $f(x_2)$ must differ in sign.

- Error. *Incorrect character input. Enter one of the following: ...* Problem Solver will give you a list of acceptable responses. Choose one of them.

- Error. *Must have positive number of selected objects.* When calculating permutations, a positive number of objects must be selected.

- Error. *N or δ is not known. First compute an estimate for N or δ.* Select [Compute δ or N] from the limit menu before requesting a graph. Then the graph will illustrate the definition of the limit.

- Error. *No data has been loaded or entered.* Remember to load data from a file or to enter it from the keyboard before trying to run statistical procedures.

- Error. *No function table has been loaded or entered.* An attempt was made to use a function table that had not yet been entered.

- Error. *No matrix has been loaded or entered.* An attempt was made to use a matrix when no matrix had been loaded from a file or entered from the keyboard.

- Error. *No model has been loaded or entered.* An attempt was made to use a linear programming model when none had been loaded from a file or entered from the keyboard.

- Error. *No output device selected. Go to the options menu, and select an output device.* Problem Solver prints output to files or directly to a printer. You have requested that a matrix be printed, but you have not told Problem Solver whether to send it to a printer or a file. All output is currently directed to the video screen.

- Error. *Number of columns can not be changed in data for statistics.* For descriptive statistics, all data must be placed in a matrix with one column. For linear regression, two columns are used. The first is for the independent variable, and the second is for the dependent variable.

- Error. *Please enter the derivative using the second menu option.* Newton's method requires both $f(x)$ and $f'(x)$. Enter $f'(x)$ using the second menu option in the equation solver menu.

- Error. *The expected input is an integer between xxxx and xxxx.* When an integer was requested, you entered one out of the range of expected responses.

- Error. *The initial guesses must not be equal.* The two initial guesses required for some of the equation solving methods must be two distinct numbers.

- Error. *The x-values must be evenly spaced.* For the requested operation, the x-values must be equally spaced.

- Error. *The x-values must be evenly spaced for the requested method. Switching to Riemann sums.* For functions defined as tables of values, all integration methods except Riemann sums require uniformly spaced x-values.

- Error. *This option is available only for functions defined as formulas.* Several options require functions defined by formulas. The study of limits and the graphing of tangent and secant lines are examples.

- Error. *Too many subintervals for selected method. Changing to xxxx subintervals.* You have requested more subintervals for integration than Problem Solver can handle.

- Error. *Two columns required for regression.* A two column data matrix is required for linear regression. The first column contains the independent variable, and the second column contains the dependent variable.

- Error. *Two or more ordered pairs have the same first element.* If you set up a function as a table of values, all the x-values must be distinct.

- Error. *Undefined function. Select enter/edit functions.* An attempt was made to use a function that had not yet been entered.

- Error. *Unevenly spaced x-values. Only the first derivative will be calculated.* For second order derivative estimates, Problem Solver requires evenly spaced x-values.

- Error. *You have only one data value.* More than one data value is needed for calculating the mean, median, and standard deviation.

- Error. *You have requested the graph of a function on an interval of the form* $[a, a]$. Select Change graph specifications to correct this. Adjust the minimum and maximum x-values to graph.

- Warning. *At least one constraint has been multiplied by* -1. If any of the constraints in a linear programming problem have negative right-hand sides, Problem Solver multiplies both sides of the constraint by -1 to produce a positive right side. This also changes the direction of the inequality sign.

- Warning. *Information line selected. No action taken.* The selected menu line gives information only. Selecting it does not cause Problem Solver to do anything.

- Warning. *Must use left-hand limit as x approaches infinity.* You requested a limit as x approaches infinity with something other than a left-hand limit. Problem Solver automatically changes to a left-hand limit.

- Warning. *Must use right-hand limit as x approaches negative infinity.* Same as last item. Problem Solver changes to right-hand limit.

- Warning. *Some of the data falls outside the intervals.* Some of the data is not included in any of the class intervals used to construct a frequency table or a histogram. This problem is most likely caused by the addition of new data points or by modifying the specifications of the frequency table.

- Warning. *The constraints have been rearranged.* Problem Solver arranges the constraints so that the less than type come first, the equality constraints come next, and the greater than type come last.

- Warning. *The objective function is unbounded.* The objective function for a linear programming problem is unbounded. This may suggest that the model was entered incorrectly.

- Warning. *The standard deviation must be positive.* You must enter a positive number for the standard deviation of a normal distribution. If you enter a negative number, it will automatically be changed to a positive number.

- Warning. *There are no feasible solutions.* There are no solutions satisfying all of the specified constraints for a linear programming problem. The model may not be correct.

C.11.6 Matrix Warnings and Error Messages

- Error. *Incompatible dimensions.* Matrix operations require compatible dimensions. To find $A + B$, the two matrices A and B must be the same size. For the product AB, the number of columns in A must equal the number of rows in B.

- Error. *Only square matrices have inverses.* An inverse was requested for a matrix that is not a square matrix.

- Error. *Requested too large a matrix.* Matrices are limited to 1000 rows and 1000 columns regardless of the amount of memory in your system.

- Error. *Singular matrix.* Singular matrices have no inverses. You will see this message if you request the inverse of a singular matrix. You will also see this message if you try to solve a system of equations whose coefficient matrix is singular. These systems have either no solution, or they have multiple solutions. Try reducing the matrix to row-echelon form for more information.

- Error. *The augmented matrix must be n by $n + 1$. Otherwise use the reduced-row method.* Both the Gaussian elimination and Gauss-Jordan methods are designed for solving a system of n linear equations in n variables. The augmented matrix for such a system has n rows and $n + 1$ columns. To solve other systems, request the row-reduced echelon form of your matrix, and find the solution from that.

C.11.7 Memory Warnings and Error Messages

The error messages in this section are very serious since a lack of memory can cause Problem Solver to behave in an erratic way or even crash. If you get any of these error messages, you should try to make more memory available to Problem Solver by removing any memory resident software that is running. In addition, if you have entered or loaded any large matrices, linear programming problems or data sets for statistics, save these in files and then remove the unneeded data from memory by selecting [Remove xxxx] from the appropriate menu.

- Error. *Can not open window.* Problem Solver can not find enough free memory to display messages on the screen.

- Error. *Insufficient memory.* Problem Solver is running short on memory.

- Error. *Insufficient memory for requested matrix.* The maximum size of a matrix that can be entered into your computer is limited by the amount of memory in your system. To see how much free memory is available, press Alt-F10 while in any menu. A matrix with m rows and n columns uses up to $8nm + 4m + 100$ bytes.

- Error. *Not enough memory to show steps.* It takes additional memory to display the steps in a linear programming problem. If Problem Solver detects this condition early enough, it simply displays this message and solves the problem without displaying the steps.

- Error. *Not enough memory to show steps. Solution may be corrupted. Begin again.* The solution to a linear programming problem may not be accurate. Select Options and then request that intermediate steps not be shown. Try once again to solve the linear programming problem.

- Error. *Out of memory. Select a smaller matrix.* Your computer does not have enough memory to handle the matrix you tried to enter.

- Error. *Out of memory. Select a smaller model.* Your computer does not have enough memory to handle the model you tried to enter.

C.12 Questions & Answers

C.12.1 Computational Problems

Q: Why does Newton's method sometimes give me incorrect answers?

A: Newton's method is the only method used by Problem Solver that requires a derivative. Be sure that you have entered the derivative correctly. If the derivative and function are both correct, and if Problem Solver reports an unexpected solution with acceptable error estimates, then your equation has multiple solutions and Problem Solver has found another solution. To find all of the solutions, try graphing the function. Estimate as closely as possible where the graph crosses the x-axis, and use this value as an initial guess. Do this for each x-intercept of your graph.

Q: I tried solving a system of equations with a matrix known to be singular. With the zero tolerance in the options menu set to 1.0e-12, Problem Solver reported a singular matrix, but when I changed the zero tolerance to 1.0e-20, an incorrect solution was given. Shouldn't a smaller zero tolerance give more accurate results?

A: A smaller zero tolerance does not necessarily lead to more accurate results. In the course of working with a matrix, computers make small errors called rounding errors. These errors are unavoidable, and they cause a lot of confusion when computers try to recognize calculations that lead to an answer of zero. Is 10^{-20} a small positive number, or is it the computer's version of 0? It is impossible to be sure. If you set the zero tolerance too small, incorrect solutions will be reported for some systems that have no solution. If you set the value too large, Problem Solver will report that some systems have no solution when they actually do have solutions. The appropriate size of the zero tolerance is not the same for all problems. You will have to experiment with this value.

C.12.2 Graphics Problems

Q: Why does the graph of the function not include the whole range of x-values represented on the x-axis?

A: The maximum and minimum x-values on the coordinate axis and the maximum and minimum x-values to graph are all adjusted separately for greater flexibility. Select [Graph functions] from the functions menu, and then select [Change graph specifications]. Using this menu, you can adjust the minimum and maximum x-values to be graphed.

Q: My graphs are displayed on the video screen, but they are not being printed at all. Is there a problem with the way I loaded Problem Solver, or is it my printer?

A: Problem Solver uses the DOS program GRAPHICS.COM to print graphs. You must run this program before running Problem Solver. In addition, your printer must be compatible with the IBM Graphics Printer.

Q: The bottom portion of my graphs are not printed. How can I print the entire graph?

A: A good way to solve persistent problems involving graphing is to start Problem Solver with the command

PS CGAHI

This reduces the screen resolution, but it does sometimes cure printing problems.

Q: Why can't I print graphs with my Hercules monochrome system?

A: Problem Solver uses the DOS program GRAPHICS.COM to print graphs, and this program supports only IBM-compatible graphics modes. A replacement program for GRAPHICS.COM that does support your system may have been included with your computer. If not, check with your computer dealer.

C.12.3 The Video Screen

Q: Why are the screens so difficult to read with my monochrome monitor?

A: Problem Solver is attempting to display color text on your monochrome screen. Start Problem Solver with the command

PS BW

Q: Why doesn't my AT&T computer use its highest resolution for graphs?

A: Problem Solver is unable to detect an AT&T graphics system. To correct this, start Problem Solver with the command

PS ATT

Q: How can I get color graphs with a CGA system?

A: Problem Solver uses the highest resolution available on CGA systems. This is the 640×200 graphics mode. Unfortunately this is a monochrome mode. The only graphics mode with colors available with a CGA system is the 320×200 four-color mode. To use this mode, start Problem Solver with the command

PS CGALO

Q: I have a VGA graphics system, but the resolution of my graphs is very poor. What is wrong?

A: The screen resolution depends on the version of DOS that you are using. If you are using a version of DOS older than 3.3, the graphs use the CGA 640×200 mode since that is the highest resolution that can be printed. For versions of DOS at least as new as 3.3, but older than 4.0, the EGA 640×350 mode is used, and for versions 4.0 and newer, the VGA 640×480 mode is used.

You can override Problem Solver's choice of a graphics mode through the use of command line parameters as discussed in Section C.3. However, if you do this, you may have problems printing graphs.

C.12.4 Miscellaneous

Q: I press F1 while entering data, but the help provided does not answer my questions. Where can I find additional help?

A: If the help provided in the data entry screen is not adequate, press the Esc key to return to the last menu. Then press F1, and read the help available. For further help, read the menu options. If one of the choices is related to your question, make that selection and press the F1 key again.

Q: I entered a function and requested its graph. The coordinate system was drawn, but my function was not graphed. Does this mean that there is an error in my function?

A: One of several things could be happening. You must set each of the 10 functions currently in memory to either *yes* or *no* for graphing. Those to be graphed should set to *yes*. Do this with the select functions menu. A function error or domain error could also prevent a function from being graphed. For example, if you request a graph of $log\, x$ for $x \in [-3, 0]$, there will be no graph since $log\, x$ is only defined for positive values of x. Syntax errors can also prevent a function from being graphed.

Q: The numbers in my matrices are too crowded. How can I spread them apart?

A: Use the options menu. Select [Format for output], and then choose a larger field for the number displays.

Q: I keep running out of memory. Is there anything that I can do to make more memory available for Problem Solver?

A: First remove any memory resident programs that you are using. This will make more memory available for Problem Solver. You can also conserve memory as you run Problem Solver. Many menus have a *remove* option, which allows you to release memory used by your data. For example, if you are solving a system of equations, remove the matrix before going on to graphing or some other Problem Solver function.

Q: What is wrong with the expression `sin x`?

A: The arguments of functions must be enclosed in parentheses. The correct expression is `sin(x)`.

Q: I am using Problem Solver on a hard disk. Must I switch to the subdirectory containing Problem Solver to run the program?

A: You can run Problem Solver from any subdirectory. The directory from which you start the program will be the default directory. This is the directory Problem Solver checks for a configuration file. If none is found, Problem Solver will use its own default values.

Q: Why do the page breaks come at the wrong places in my printed reports?

A: You can adjust the number of lines per page that are printed using the format option in the options menu. Problem Solver keeps track of how many lines it has printed on any page. If you should ever take the printer off-line and press the form feed button, Problem Solver would not know that you did this and would lose track of the number of printed lines. Be sure to let Problem Solver advance the paper until you are completely through printing. If you want to control when the printer advances to a new page, select Options menu and then select the User control over page breaks.

Printing graphs can sometimes change the line spacing on a printer. Problem Solver uses control codes to reset the line spacing to six lines per inch for Epson and IBM compabible printers. These codes are found in the format menu which is accessible from within the options menu. For other printers you should change these codes to the ones appropriate for your printer. These codes can be found in your printer manual.

Q: I changed some of the selections in the options menu and saved them. Now I would like to use the original program default values. How can I change back?

A: From the DOS command line delete the file PSOPTION.CTL. If Problem Solver can not find this file, it will use its own iternal default values.

Q: Is there a way to create a document, edit it, and then print it all at once?

A: Use the options menu and select file for the output device. All of your program output will then be written to a file. You can control the line length and other format features using the format option in the options menu. After you have completed your calculations, exit Problem Solver, and load the output file into a word processor for whatever editing is necessary.

The graphs must be printed directly. They cannot be written to a file for later printing.

Q: I selected the printer as the output device, but nothing happened. How can I print my results?

A: Select the printer as the output device *before* doing your calculations. Results calculated after this selection will be printed. For example, you might experiment with a graph until it looks right on the screen. Then select the printer as your output device and again draw the graph on the screen. This time it will be printed.

Appendix D

The Simplex Algorithm

Problem Solver uses the simplex algorithm to solve linear programming problems. We shall illustrate this process with an example: Find the maximum value of

$$Z = 2x_1 + x_2$$

subject to the functional constraints

$$\begin{aligned} x_2 &\leq 5 \\ x_1 + 2x_2 &\leq 12 \\ 4x_1 + x_2 &\leq 20 \end{aligned}$$

and the nonnegativity constraints

$$x_1 \geq 0, x_2 \geq 0.$$

Since it is easier to work with equations than with inequalities, we shall rewrite the functional constraints as equations by introducing *slack variables*. The constraint

$$x_2 \leq 5$$

is equivalent to the combination of the equation

$$x_2 + x_3 = 5$$

and the nonnegativity constraint $x_3 \geq 0$. Note that x_3 represents the difference between 5 and x_2, the right side and the left side of the first functional constraint. In the same manner we introduce the slack variables x_4 and x_5 into the last two functional constraints. Including the equation for Z in the constraints, our new problem becomes: Find the maximum value of Z subject to the constraints

$$\begin{array}{ccccccccccc} Z & - & 2x_1 & - & x_2 & & & & & = & 0 \\ & & & & x_2 & + & x_3 & & & = & 5 \\ & & x_1 & + & 2x_2 & + & & x_4 & & = & 12 \\ & & 4x_1 & + & x_2 & + & & & x_5 & = & 20 \end{array}$$

$$x_i \geq 0 \text{ for } i = 1, 2, \ldots 5.$$

The functional constraints have been converted to a system of three equations in five variables. Don't count the Z equation; it did not come from a constraint. The standard method for solving such a system is to assign values to two of the variables and solve for the remaining three. We shall set x_1 and x_2 equal to 0. These variables are called *nonbasic* variables. We then solve for the remaining three variables x_3, x_4, and x_5. These are called *basic* variables. When we set two (the number of original variables or decision variables) of the variables equal to zero and solve for the remaining three (the number of constraints) we get a *basic* solution. It can be shown that if a maximum or minimum value of Z exists, then Z will have this optimal value at a basic solution. The simplex algorithm is an iterative algebraic procedure for systematically stepping through the basic solutions until an optimal solution is found.

Because of the form of the equations, it is easy to see that if $x_1 = x_2 = 0$, then $x_3 = 5$, $x_4 = 12$, and $x_5 = 20$. That is, our first basic solution is $(0, 0, 5, 12, 20)$. Notice the current form of Z. Solving the first equation for Z yields

$$Z = 2x_1 + x_2 + 0x_3 + 0x_4 + 0x_5.$$

Since Z is written in terms of the nonbasic variables all of which have value 0, the value of Z at our first basic solution is 0. The only nonzero variables are x_3, x_4, and x_5, and they have 0 coefficients in the expression for Z.

In moving from our first basic solution to the next, we would like to increase the value of Z as rapidly as possible. Choosing a new basic solution amounts to selecting which two variables will be set equal to 0 before solving for the remaining 3. This means that we could increase either x_1 or x_2 from 0 to some positive value. Since x_1 has a coefficient of 2 in the expression for Z, and since x_2 has a coefficient of 1, it appears that we could increase Z faster by increasing x_1 from 0 to some positive value. After all, each unit of increase in x_1 increases Z by 2 units whereas each unit of increase in x_2 increases Z by only 1 unit. Thus x_1 becomes a basic variable in our next solution. Since x_1 enters the set of basic variables, it is called the *entering* variable.

To make room for x_1 in the set of basic variables, one of the old basic variables, x_3, x_4, or x_5 will have to become nonbasic. It is called the *leaving* variable. To select the leaving variable, consider the constraint equations. We can find the maximum value that x_1 can have by setting all other variables equal to zero. From the first constraint

$$0x_1 \leq 3.$$

Since this inequality is true for any value of x_1, it imposes no restrictions on the maximum size of x_1. From the second constraint

$$x_1 \leq 12,$$

and from the last constraint

$$4x_1 \leq 20.$$

For the last two inequalities to both be satisfied, we need $x_1 \leq 12$ and $x_1 \leq 20/4$. Since $5 < 12$, both inequalities will be satisfied as long as $x_1 \leq 5$. Thus 5 is the largest value that x_1 can have if all of the constraints are to be satisfied. We shall increase x_1 to 5. The last constraint determines how large x_1 can be. Written in the form of an equation with the slack variable included, this constraint is

$$4x_1 + x_2 + x_5 = 20.$$

Increasing x_1 to 5 requires that both x_2 and x_5 be 0. Since x_2 is nonbasic it is already 0. Since x_5 must be reduced to 0, we make it nonbasic. The leaving variable is x_5.

We now set x_3 and x_5 equal to 0 and solve for the remaining variables. A slight variation of the Gauss-Jordan method can be used to do this. The objective is to rewrite the system of equations in the simple form of the original system in which each basic variable appeared in only one equation and with coefficient 1. In addition, Z was expressed in terms of nonbasic variables only. We want each basic variable to have a 0 coefficient in each equation except for one. The coefficient of the basic variable in that equation is to be 1.

Adding 1/2 times the last equation to the first equation yields

$$Z + 0x_1 - 0.5x_2 + 0x_3 + 0x_4 + 0.5x_5 = 10.$$

Since the coefficients of the basic variables are 0, Z is again written in terms of nonbasic variables. The left side of the above equation simplifies to Z, and we immediately see the Z has increased from 0 to 10 in our first step.

The second equation already has 0 as the coefficient of x_1 so it need not be changed. Adding $-1/2$ times the last equation to the third equation leads to the equation

$$0x_1 + 1.75x_2 + 0x_3 + x_4 - 0.25x_5 = 7.$$

Dividing the last equation by 4 row produces

$$1x_1 + 0.25x_2 + 0x_3 + 0x_4 + 0.25x_5 = 5.$$

We now have the new system

$$\begin{array}{rcrcrcrcrcrcr} Z & + & 0x_1 & - & 0.5x_2 & + & 0x_3 & + & 0x_4 & + & 0.5x_5 & = & 10 \\ & & 0x_1 & + & x_2 & + & x_3 & + & 0x_4 & + & 0x_5 & = & 5 \\ & & 0x_1 & + & 1.75x_2 & + & 0x_3 & + & x_4 & - & 0.25x_5 & = & 7 \\ & & x_1 & + & 0.25x_2 & + & 0x_3 & + & 0x_4 & + & 0.25x_5 & = & 5 \end{array}$$

Look at the columns corresponding to the basic variables x_1, x_3, and x_4. Since all other variables are set equal to 0, we need consider only these columns when solving for the values of the basic variables. The solutions can now be read from the right-hand sides of the equations. If $x_2 = 0$ and $x_5 = 0$, then $x_3 = 5$, $x_4 = 7$ and $x_1 = 5$.

We now repeat the process by again selecting an entering variable. Solving the first equation for Z yields

$$Z = 0x_1 + 0.5x_2 + 0x_3 + 0x_4 - 0.5x_5 + 10.$$

We can increase either of the nonbasic variables x_2 or x_5 from 0 to some positive value. Note that increasing x_5 will cause Z to decrease. Increasing x_2 will increase the value of Z, so x_2 will be our entering variable. Let us review the process for selecting this entering variable. After solving for Z we select the variable with the largest positive coefficient. This will cause Z to increase as rapidly as possible. Once this is understood, we no longer need to solve for Z. We can look at the equation in the form

$$Z + 0x_1 - 0.5x_2 + 0x_3 + 0x_4 + 0.5x_5 = 10$$

and pick the negative coefficient with largest magnitude—the one that would become the largest positive coefficient if we were to solve for Z.

To select the leaving variable we examine the constraints to see how large the entering variable can become. From the first constraint $x_2 \leq 5$. The other constraints require $1.75x_2 \leq 7$ and $0.25x_2 \leq 5$. Solving for x_2 in each of these we get $x_2 \leq 5$, $x_2 \leq 7/1.75$, and $x_2 \leq 5/0.25$. To satisfy all of the inequalities simultaneously and at the same time increase x_2 as much as possible, we set x_2 equal to the smallest of 5, 7/1.75, and 5/0.25. Consequently $x_2 = 7/1.75$. Since the third equation determined the maximum value of x_2 and since x_4 is the basic variable in that constraint equation, x_4 is the leaving variable. Now x_4 and x_5 are the nonbasic variables. We shall set $x_4 = 0$ and $x_5 = 0$ and solve for the remaining (basic) variables.

We again use row operations to express Z in terms of nonbasic variables and each equation in terms of a single basic variable. What this means is that the coefficient of the entering variable is to be 1 in the third equation (the one from which the leaving variable was selected) and 0 in all other equations. We accomplish this be replacing the first equation with the sum of the first equation and 0.5/1.75 times the third equation. Next we replace the second equation by the sum of itself and $-1/1.75$ times the third equation. The last equation is replaced by the sum of itself and $-0.25/1.75$ times the third equation. Finally the third equation is replaced by 1/1.75 times itself. The resulting set of equations follows:

$$\begin{array}{ccccccccccccc}
Z & + & 0x_1 & + & 0x_2 & + & 0x_3 & + & 0.286x_4 & + & 0.429x_5 & = & 12 \\
 & & 0x_1 & + & 0x_2 & + & x_3 & - & 0.571x_4 & + & 0.143x_5 & = & 1 \\
 & & 0x_1 & + & x_2 & + & 0x_3 & + & 0.571x_4 & - & 0.143x_5 & = & 4 \\
 & & x_1 & + & 0x_2 & + & 0x_3 & - & 0.143x_4 & + & 0.286x_5 & = & 4
\end{array}$$

In this form we see that Z has increased to 12. We also see that if $x_4 = 0$ and $x_5 = 0$, then $x_3 = 1$, $x_2 = 4$, and $x_1 = 4$.

There are no negative coefficients in the equation defining Z. This means that we cannot increase Z by increasing one of the nonbasic variables. Our current solution is optimal. The maximum value of Z is 12, and the other variables have values $x_1 = 4$, $x_2 = 4$, $x_3 = 1$, $x_4 = 0$, and $x_5 = 0$. Of course, the first two variables are the only ones that were present in the original problem. Since $x_3 = 1$ and x_3 is the slack variable for the first constraint, there is some slack in that constraint. At the optimal solution, the difference between the right side and left side of the original constraint $x_1 \leq 5$ is 1. The fact that $x_4 = x_5 = 0$ tells us that there is no slack in the other constraints. That is, at the optimal solution $x_1 + 2x_2 = 12$ and $4x_1 + x_2 = 20$.

Just as systems of equations are more easily handled with matrices, linear programming problems are more easily handled with tableaux or tables of coefficients. The first system with slack variables can be represented by the following tableau:

Basic	Coefficient of						Right
variable	Z	x_1	x_2	x_3	x_4	x_5	side
Z	1	−2	−1	0	0	0	0
x_3	0	0	1	1	0	0	5
x_4	0	1	2	0	1	0	12
x_5	0	4	1	0	0	1	20

The current basic solution is $(x_1, x_2, x_3, x_4, x_5) = (0, 0, 5, 12, 20)$, and the current value of Z is 0. Even though Z is not a basic variable, we include it in the left column to emphasize the fact that we solve for Z along with the basic variables after setting the nonbasic variables equal to 0. Because of the form of the tableau, we can solve for these variables by inspection. The right column of the tableau gives the values of the variables in the left column.

Looking at the first row, we see that x_1 has the negative coefficient with largest magnitude. Thus x_1 is the entering variable. To select the leaving variable we examine the ratios formed by dividing the values on the right sides of the equations by the positive coefficient of the entering variable. These ratios are (omitting the first since the coefficient of x_1 in the second row is 0) 12/1, and 20/4. The minimum ratio determines the leaving variable. It is x_5.

We build the next tableau from the first by using row operations to produce zeros throughout the column containing the entering variable except for the coefficient of the entering variable in the row containing the leaving variable. That coefficient is made equal to 1 by a division if necessary. We replace the first row by the sum of itself and 1/2 times row 4, the third row by the sum of itself and −1/4 times row 4, and row 4 by 1/4 times itself. The new tableau is

Basic	Coefficient of						Right
variable	Z	x_1	x_2	x_3	x_4	x_5	side
Z	1	0	−0.5	0	0	0.5	10
x_3	0	0	1	1	0	0	5
x_4	0	0	1.75	0	1	−0.25	7
x_1	0	1	0.25	0	0	0.25	5

The current basic solution is $(x_1, x_2, x_3, x_4, x_5) = (5, 0, 5, 7, 0)$, and the current value of Z is 10. Notice that these values can easily be read off the right side of the tableau. One advantage of the tableau form is that we have used the left column to identify the basic variable for each row. Its value is the corresponding value in the last column. Of course the other variables are nonbasic and therefore 0.

Looking at the first row, we see that x_2 has the negative coefficient with largest magnitude, and it is the entering variable. To select the leaving variable we examine the ratios 5/1, 7/1.75, and 5/0.25. The smallest ratio is 7/1.75, so x_4 is the leaving variable.

We use row operations to put the column corresponding to the entering variable x_2 in the proper form. That is, all coefficients are to be 0 except for the row corresponding to the entering variable. We replace the first row by the sum of itself and 0.5/1.75 times the third row. The second row is replaced by the sum of itself and $-1/1.75$ times the third row, the fourth row is replaced by the sum of itself and $-0.25/1.75$ times the third row. Finally the third row is replaced by 1/1.75 times itself. This produces the tableau

Basic	Coefficient of						Right
variable	Z	x_1	x_2	x_3	x_4	x_5	side
Z	1	0	0	0	0.286	0.429	12
x_3	0	0	0	1	−0.571	0.143	1
x_2	0	0	1	0	0.571	−0.143	4
x_1	0	1	0	0	−0.143	0.286	4

Since all of the coefficients in the first row are nonnegative, there is no entering variable. The current solution is optimal. That solution is

$$Z = 12,\ x_1 = 4,\ x_2 = 4,\ x_3 = 1,\ x_4 = 0,\ x_5 = 0.$$

Notice that the second column, which gives the coefficient of Z in each equation never changes during the row operations. To save space Problem Solver omits this column in its displays.

The above process will work on any linear programming problem of the maximization type with less-than-or-equal-to constraints, with nonnegative right-hand sides in the constraints, and with the usual nonnegativity constraints. There are variations of this method for handling equality and greater-than-or-equal-to type constraints. Problem Solver uses the two-phase version of the simplex algorithm in these cases. The interested reader is encouraged to refer to the listed references for additional information. Problem Solver closely follows the algorithm presented in the first reference.

1. Frederick S. Hillier and Gerald J. Liebermann, *Introduction to Operations Research*, 1986, Holden Day.

2. Wayne L. Winston, *Operations Research: Applications and Algorithms*, 1987, PWS Publishers.

Index